BIODIVERSITY AND ECOSYSTEM CONSERVATION

BIODIVERSITY AND ECOSYSTEM CONSERVATION

Editor
ASHISH DUTTA

KALPAZ PUBLICATIONS
DELHI-110052

BIODIVERSITY AND ECOSYSTEM CONSERVATION
(Environment, Ecology)

ISBN: 81-7835-058-0

Published 2016 in India by
Kalpaz Publications
C-30, Satyawati Nagar
Ashok Vihar Phase-III
Delhi-110052
E-Mail: kalpaz@hotmail.com

Laser Type setting at : A.S. Printers
Printed at : Mehra Offset Printers, New Delhi

Preface

Biodiversity is the variety of the world's living species, including their genetic diversity and the communities and ecosystems that they form. It is the result of two opposite actions: the processes that produce new genotypes, new varieties, and new species and the processes that eliminate mutations, variants, and species from the system. Natural selection is primarily responsible for the reduction of biodiversity; it acts through differential reproduction and differential mortality.

Biogeographical studies have shown that ecosystem diversity varies greatly from one region to another. Tropical forests, coral reefs, and deep benthic communities are very rich in .species: deserts and high-latitude areas are species poor. Many studies have shown that species abundance changes naturally along latitudinal and humidity gradients. A global survey of diversity is essential to understanding the distribution of life forms on the planet.

Today, this diversity of life is facing serious threats. Scientists concur that, at current rates of tropical forest loss and disturbance, roughly 5 to 10 per cent of all tropical forest species will be lost per decade during the next quarter century. The extinction crisis is not restricted to tropical forests. Freshwater habitats are being altered dramatically as rivers are impounded and exotic species introduced.

The complex threats to biological diversity call for a wide range of responses from many private and public sectors, and the mix of these responses must be adjusted to suit local conditions. Because government policies often are responsible for depleting biological resources, policy changes usually are a necessary first step toward conservation. National policies that deal directly with wildlands management or forestry or that influence resource use indirectly by affecting, for example, land tenure, rural development, family planning, food prices, pesticide use, or fuel consumption can have significant impacts on the conservation of biodiversity.

This book complies succinct and authoritative information about biodiversity. It will certainly prove an asset and reference tool to all concerned.

Editor

Contents

1

Introduction

Today, physicists believe that nature is complex, not simple, and that the Newtonian view is insufficient as a general explanation of how the universe functions. The universe, physicists are finding, is not in balance. Disturbances and irregularities of all sorts are no longer seen as aberrations but as integral parts of nature.

Physicists are also discovering that at the very origin of the cosmos, at the so-called big bang, a singular, irreversible, and complex universe was produced. Similarly, ecologists have been observing and documenting that most, if not all, ecosystems are not in balance. But why should scientists be concerned with whether an ecosystem is in equilibrium? The answer to this question is crucial for understanding biodiversity.

Systems not in equilibrium behave very differently than do systems in equilibrium. Their behaviour can even appear strange and mysterious. For example, when the source of a disturbance is removed from a system that is near equilibrium, that system is expected to return to its previous state. When, in an undisturbed forest, a gap is created by landslide or a falling tree, new seedlings grow in the gap and restore the equilibrium. After a few years, it is almost impossible to tell that there once was a gap. But when the source of change is removed from systems that are not in equilibrium, they do not return to equilibrium. Instead, they adopt a new state.

For example, when land used for agriculture in the Amazon forest is abandoned, it grows into a grassland or a savanna and does not return to the original forest. Natural disturbances are a necessary part of the forest ecosystem's function. For example without gaps creating disturbances, most forests cannot renew themselves. And fires, storms, and hurricanes are and have been a part of life since the beginning of the planet; without them, ecosystems could not function properly.

Knowing about the features of non-equilibrium systems is important for proper ecosystem management. Because a non-equilibrium ecosystem does not necessarily revert to its previous state when the convulsion is removed it is often impossible for humans to return ecosystems to their original condition. Therefore, one set of management guidelines will not suffice for all situations.

The previous history of changes and the present disturbance regime will determine the consequences of any management. There are two conditions that cause population diversity.

- New genotypes are constantly cropping up in a population through mutation, recombination, and related genetic phenomena and through immigration of individuals, their gametes, or their propagules.
- Diversity in the population is eliminated by natural selection and lost through emigration of individuals.· Every genetic variation, from gene

mutations to entire species, will disappear eventually.

This loss can be a very fast process, or the variants can survive for a long time. Species that have survived for extended periods include horseshoe crabs, which have been around for 200 million years, and cockroaches, which originated even earlier, in the Carboniferous period.

The speed at which new variations originate in relation to the rate at which they are eliminate determines the actual diversity of the system. In the process of gene mutation, all heritable diversity ultimately arises at the molecular level. Gene mutations are chemical changes that take place in the composition of the DNA (deoxyribonucleic acid) molecule, the chemical substance responsible for heredity DNA is remarkable in that it shows complex chemical behaviour not expected from systems in equilibrium.

The most remarkable characteristic of DNA is its ability to regenerate itself, its "autocatalytic" behaviour. The DNA molecule consists of four kinds of repeating units called uncleotides, which can be arranged in any sequential order. DNA has the curious ability to maintain its physicochemical integrity regardless of the order of the four nucleotides. That is, any DNA molecule behaves like any other of the same length, regardless of the nucleotide content. This characteristic of DNA makes life feasible because if only one arrangement were possible, or if the chemical stability of one arrangement were significantly different from that others, then all DNA molecules would be alike.

The order of the nucleotides uniquely determines the characteristics of the chemical products made by DNA. But the order of the nucleotides in a DNA molecule can change, and these changes are called mutations. The regular appearance of *mutations* gives life its diversity. Mutated molecules of DNA reproduce their changes and make *modified enzymes*, which, in turn, make *altered cells*.

Mutated cells result in *modified organisms*. Thus, new mutations permit the evolution of new characteristics by changing the *structure* and /or *function* of enzymes and other *proteins*. The characteristics of the enzymes and proteins are determined by those of the DNA molecule, but not the reverse — the "information" flows only in one direction. This flow contrasts with that of an ordinary chemical reaction, in which there are forward and backward reactions.

The appearance of mutations is a *random* process. There is no way to anticipate or predict what mutation is going to occur, only that some will. It is natural *selection*, however, that determines whether new mutations get established or are eliminated. The processes of mutation and selection can be looked at in two ways. Mutation and selection might be random, independent processes that are not influenced by the characteristics of the system, in which case no active process of diversity maintenance exists.

On the other hand, a certain degree of variety may be required for living systems to function properly. If so, there may be *system feedbacks* that modify the rates of mutation and selection. Below a certain threshold, a system may collapse because of insufficient variety. According to this last view, diversity is actively maintained in a system.

Mutation and natural selection generate genetic and morphological variation within a lineage. Yet what is most distinctive in a community is the coexistence of different species, each acting independently and not sharing its DNA or mutations. How, then, does the great variety of species originate? Speciation is the process that separates genetic variatión into distinct units, or species. In speciation, the original population of organisms with similar genes, called a gene pool, is divided into two or more gene pools. Each of these new gene pools then acquires a unique set of characteristics (cellular, tissue, organ, and organismic) through mutation and selection. Speciation follows most commonly from the physical division of a gene pool.

The separation inhibits interbreeding between individuals in the two populations. In time, one or both populations change enough to prohibit interbreeding. Because the fate of every species is to become extinct eventually, there must be an influx of new species for life to continue. If, presently, there are about 10 million species, and, on average, each species has a life span of 1 million years, then an average of 10 new species must originate each year. On the other hand, if the life span of a species is only 100,000 years (as many biologists argue), then 100 new species must originate each year.

Scientists are often aware of a species' extinction, but the appearance of new species is difficult to observe. Because most species are insects and other small invertebrates, most new species fall into these two poorly studied groups. Thus, there are few well-documented cases of speciation. Certain individuals in a population survive and reproduce, while others die without leaving offspring. To understand this phenomenon, scientists must first realize that there usually is not enough of some environmental resources (food, water, minerals) to go around. The availability of resources in the environment limits the rate at which individual organisms can reproduce and their ability to survive. Nucleic acid replication qnd cell, tissue, and organism growth and reproduction all require energy and materials, of which there are only limited amounts.

Individuals that are efficient at harvesting resources have a higher probability of reproducing and surviving than do those that are inefficient. Mutations in DNA provide species with a constant input of physical and chemical variations, some of which improve the efficiency with which the species selects and harvests resources. New mutations may increase the efficiency of protein synthesis at the molecular level, or they may improve the water uptake by a plant at the organism level. Mutations may even increase the efficiency with which a mutualistic association between a plant and its specific pollinator develops at the community level. DNA molecules that are dissimilar in the order of their nucleotides usually do not have characteristics that directly increase or decrease their own survival or reproductive capacity.

Given the appropriate environment and resources, all DNA molecules reproduce at the same rate despite their nucleotide composition. DNA molecules manifest their distinctive survival rates through their effects on the living organisms in which they are embedded. These organisms, be they single cells living in pure culture or multicellular organisms living in complex communities, will inevitable differ in their ability to garner resources. Therefore, they will vary in their survival and reproductive capacities and, consequently, so will the DNA in their cells. Natural selection is an optimizing process. It is process that favours efficiency and produces adaptation. It is not a process that involves chance, like the appearance of mutations, but it must operate with the elements that it has. Thus, natural selection never makes a "perfect" organism. Because the environment is constantly changing, a perfect organism, if it could be produced, would soon cease to be perfect as the environment changed. Such systems, in which the optimum state is favoured but cannot be attained or maintained, are called frustrated systems. Diversity is the result of two opposite actions: the processes that produce new genotypes, new varieties, and new species and the processes that eliminate mutations, variants, and species from the system.

Natural selection is primarily responsible for the reduction of biodiversity; it acts through differential reproduction and differential mortality. In order words, the probability of reproducing and the probability of dying are not the same for all organisms. Rather, reproduction and death are related to the organism's characteristics as well as to the environment in which that organism lives. How is species diversity linked to ecosystem structure and function, to the different ways in which organisms and population interact with each other, and to those community properties that emerge from these interactions?

Communities have structures and properties not possessed by the populations within them that are called emergent properties, which include trophic structure, stability, guild structure, and successional stages. Many scientists believe that species diversity is essential for the proper functioning of communities and for the emergence of community-level properties. Just as many different DNA-encoded enzymes are needed for a complex organism to function properly, so, scientists believe, are many kinds of species necessary to maintain community structure. But is any diversity sufficient, or are specific mixes of species necessary for the communities and ecosystems of function? This is a very old question in ecology, and two opposing views exist. One view is that a community is formed by the species that happened to arrive first — that the mix of species in a community is a matter of chance: "The vegetation of an area is merely the resultant of two factors, the fluctuating and fortuitous immigration of plants, and an equally fluctuating and variable environment." According to the opposing view, "In any fairly limited area, only a fraction of the forms that could theoretically do so actually from a community at any one time.... The community really is an organized community in that it has a 'limited membership.' "

Are these two very different views really irreconcilable? Not necessarily, if a community is defined as the sum of all the plants and animals that grow together in an area, certain patterns can be observed: Areas with more than 1,000 millimeters of evenly distributed rainfall always contain a woodland; conifers prevail in areas of extremely low winter temperatures; trees in warm climates have broad leaves; and succulent plants are found in dry climates. The same is true for animals. The mix of species encountered in a community is not a random sample of all plants and animals in the world. Yet, under close scrutiny, differences that are difficult to explain can be found between similar localities. Ecologists cannot predict what species will occur in a given climate and soil type.

Also, there have been cases where dominant species, such as the American chestnut, have been lost, with apparently little effect in the overall working of the biological community. The community behaves in a "frustrated" fashion. Two types of processes are at work and neither is dominant. First, there is natural selection. Species in the community are constantly evolving to increase their ability to withstand the rigours of the environment. This process accounts for cold-hardy species near the popes, for grasseating ruminants in savannas, and for fruiteating bats in forests. The individuals that escape their predators and survive other environmental challenges pass on their characteristics to their offspring. Second, there is chance. When the climate changes, characteristics that were once advantageous can become a burden. Diseases might be introduced, such as Dutch Elm disease, for which local species have no defense. Hurricanes, tornadoes, floods, and droughts are unpredictable and their frequency changes over time.

Chance also plays a role in dispersal. For instances, a species might be absent from a community purely by chance. Thus, both natural selection and chance result in the steady coming and going of species through immigration, emigration extinction, and gene mutation. A large number of additional unanswered questions exist regarding the role of biodiversity in communities.

For example, can more and more species be packed into a community, or are there upper and lower limits? A related question concerns optimal levels of diversity and the factors that control them. Other issues include the role of different individuals within a population, different populations within a species, and so on. Do the members of the community collaborate in the efficient use of energy and resources? In a variable environment, the existence of individuals with different characteristics may increase the ways in which the population can respond to change. For example, if all plants of a species had similar water requirements, all of them would suffer water stress in any year that was drier than normal, and such periods would result in significantly reduced seed production. But if there is genetic variation, some individuals might perform above average each year within certain limits of environmental variation.

Therefore, seed production would be satisfactory in both wet and dry years. Moreover, genetically adaptable organisms should survive in more variable environments than do genetically uniform populations.

Experience with crops shows that highly productive but genetically uniform varieties have more restricted environmental requirements than do less productive but more variable varieties. Also, plantations formed by uniform varieties are more susceptible to pest and disease outbreaks. So, variability within a species seems to be important for long-term survival. Does the same hold for a community, which is formed of species that live together but do not have a common gene pool? Communities with high species diversity may also cope with long-term environmental fluctuates better than do communities with few species. However, the evidence is contradictory.

Terrestrial communities in the climatically variable midlatitudes are less diverse than tropical communities in more uniform environments. Also, deep benthic communities are among the most variable communities anywhere though they exist in possibly the most equitable environment on the planet. According to modern ecological theory; every species in a community occupies a singular ecological niche, or a role in the environment.

A niche is thought of as a multidimensional space where each dimension is a characteristic of a species. Because every species has at least one physical or behavioural characteristic that separates it from other species, every species has a unique niche. In principle, niche theory supplies the theoretical framework to explain the number and types of species in a community. Niche theory predicts that communities that vary in total resources or in the quality of those resources will contain more niches and, therefore, more species. Nevertheless, resources are sometimes partitioned differently in communities with very similar resources.

In other words, communities may be have large or small niches, which result in a few "generalist" species (those with board niches) or many "specialist" species (those with small niches). Finally, communities with similar resources and niche sizes may still vary in the number of species because of different degrees of niche overlap. Niche overlap is the degree to which two species exploit the same resource. Communities differ in niche size and overlap for a variety of reasons, such as differences in climatic stability and predictability, spatial heterogeneity, primary productivity, computation predation, and degree of disturbance. Thus, niche theory is unable to predict the patterns of species diversity.

Using niche theory to understand why communities vary in number of species has proven very difficult in practice. Many studies have shown very elegantly how species partition various resources in space and in time, but it has proven very hard to show the precise contribution of each factor.

A key question is whether two species can have the same niche. In other words, is there species redundancy in a community? Present theory predicts that two species with identical niches cannot coexist, and empirical studies seem to verify this. Still, the question of how similar two coexistent species can be remains to be answered.

Additional unanswered questions concern the role of diversity in food webs. All organisms get their energy from the sun, either directly or indirectly. At the simplest level, cells of the green tissues of plants and some micro-organisms capture the sun's energy. Organisms also need specific chemicals from the environment to develop. For instance, all plants obtain the materials for development from soil or water and the air. With these materials and energy from the sun, plants create energy-rich,complex chemical compounds, such as carbohydrates, cellulose,starch, proteins, fats, and wood.

Animals ingest and extract the energy from these compounds and incorporate some of them into their bodies. Decomposers reduce them to simple ions that eventually are returned to the soil or water. In a cyclic fashion, then, energy and chemical substances move from the environment to plants, to animals, to decomposing microorganisms, and back to the environment. This trophic diversity, or "food web", is essential for the proper operation of ecosystems. Many studies have shown that the trophic structure of ecosystems is complex and varies from one ecosystem to another.

Understanding how food webs function is essential for understanding the place of humans in the ecosystem,[19] because humans also derive their energy and materials from plants and animals. Moreover, interpretation of food webs is necessary to understand the effect of toxic compounds released into the environment and the effect of the introduction of species into new areas. Are diverse ecosystems with complex food webs more stable than simple ecosystems?

Stability, in this instance, refers to the ability of an ecosystem to withstand disturbances, including those produced by people. Scientists used to believe that simple systems are less stable than complex ones. Clearly, there is a threshold of diversity below which most ecosystems cannot function. For example, no ecosystem can function without some plants and decomposers. Without plants, solar energy cannot be captured, and without decomposers, substances are not returned to the environment. But, is there a threshold below which present complex ecosystems lose their stability? How many different sorts of plants and animals are needed? and, if there are many of each sort, does the ecosystem become more stable?

Productivity, or the total quantity of energy captured and biomass produced by an ecosystem in a year, is related to the issue of stability. Are ecosystems that are formed by many species more productive than those with few species? It would appear that simple systems, both natural (*Spartina* marshes and high-latitude marine systems) and artificial (agricultural systems), are more productive than diverse ones.

However, some scientists fear that these systems are not stable and have argued that society's increasing reliance on the productivity of these simple systems could eventually put humans in jeopardy. Apart from the ethical and aesthetic reasons, is there any reason to fear that human survival is at risk if biodiversity is not preserved? Can humans exist surrounded only by agricultural fields, planted forests, and the like? As long as the environment does not change very much, ecosystems can apparently lose many of their rare species without any visible effects.

In some cases, common animal species, such as passenger pigeons, have disappeared or been drastically reduced without endangering human survival. Yet, environments inevitably changes, and sometimes drastically. It is then that the role of obscure species sometimes becomes very important. When dinosaurs became extinct, an obscure group, primitive mammals, suddenly evolved into the dominant role they still play. And now, humans seem on the verge of a drastic changes, one of their own making—the result of global warming induced by increased carbon dioxide in the atmosphere.

Predicate climatic changes pose a series of specific questions relating to biodiversity for which there are presently no answers. What is the impact of elevated atmospheric concentrations of carbon dioxide on tree physiology and forest composition? How does elevated carbon dioxide affect soil organisms and soil nutrients? How is climatic change going to affect insect populations, fire frequency, and wind storm? Coastal zones, in particular, are likely to reflect phenomena occurring over a larger geographical area.

Already, the North Sea coast is showing the effects of the industrial and agricultural activity of central Europe, which is hundreds of mile away. The effects of industrialization may be catastrophic for many species, which may become extinct. But a few species, and there is no way of predicting which ones, may thrive in the new environment. Therein lies the scientific importance of biodiversity: It increases the likelihood that at least some species will survive and give rise to new lineages that will replenish the Earth's biodiversity. A global survey of diversity is essential to understanding the distribution of life forms on the planet. Biogeographical studies have shown that ecosystem diversity varies greatly from one region to another.

Tropical forests, coral reefs, and deep benthic communities are very rich in species: deserts and high-latitude areas are species poor. Many studies have shown that species abundance changes naturally along latitudinal and humidity gradients. An unprecedented transformation of the natural landscapes of the world is now occurring. Yet, different areas are being transformed at different rates. Some regions, such as the Mediterranean basin, have been altered by people for a long time. Other regions have been influenced only moderately by humans. Also, studies show that the response to disturbance varies greatly from one ecosystem to another. Therefore, it is important to determine how different biomes respond to human disturbance and to document the resulting change in biodiversity and ecosystem function. Classical biogeography concentrated on the study of species ranges and the causes of their distribution. Given scientists' uncertainty about the number of species and the large number of undescribed species, new and more efficient approaches must be developed to complement past studies.

These new approaches should take advantage of remote sensing technology and produce a geographic information system with statistical and analytical capabilities. However, because satellites cannot detect species diversity, new measures that are indicative of diversity and are more easily obtained by remote sensing must be used. For example, satellites can detect standing biomass and productivity, and research should help to establish the relationship between biodiversity and these measures. Other problems to be studies include the effect that habitat fragmentation has on species ranges and on the probability of extinction and speciation; whether fragmented habitats are more prone to invasion; how invasions influence biodiversity; and the effect of increased geographical isolation on populations and species.

Also, research should be conducted to develop a methodology for the systematic comparison of biodiversity across regions. To this end, a biogeographical information system for assessing diversity might use satellite information to identify assemblages of species at different geographical scales for long-term monitoring.

Until now, research in biodiversity has been undertaken by taxonomists at such institutions as the

Smithsonian Institution, the Harvard University, Herbaria and Musea, the Missouri Botanical Garden, the Royal Botanical Gardens in Kew, the Musea d'Histoire Naturelle in Paris, and the Komarov Institute in Leningrad, which are located for the most part in the north temperate zone, while biodiversity is found mostly in tropical countries.

Taxonomic institutions in both the north and the south have been in relative decline over the last 50 years. Staffs have not increased or have increased only slightly, and resources allocated to taxonomy have not grown in the same proportion as those devoted to other areas of biology. Consequently, at a time when the conceptual and practical importance of biodiversity is becoming clear, there is a shortage of trained personnel and a lack of necessary funds. Taxonomists alone cannot answer the questions that have been raised by the massive transformation of the Earth.

A collaborative effort by many kinds of biologists, including geneticists, physiologists, and ecologists, is required. As new techniques of genome analysis become routine, even molecular biologists may contribute to identifying genetic variation and its biological role. Both governmental and non-governmental institutions should collaborate to advance scientific understanding of the significance of biodiversity, and many more resources will have to be made available to fund the collaborative research.

The International Union of Biological Sciences, the Scientific Committee on Problems of the Environment, and the Man and the Biosphere Programme of the United Nations Educational, Scientific and Cultural Organization are spearheading research in biodiversity. These organizations are in close contact with the International Geosphere-Biosphere Programme and the United Nations

Environmental Programme, to minimize overlap and to coordinate activities.

Many national institutions also have active programmes of research in biodiversity, but most of these efforts suffer from inadequate funding. Biodiversity research and conservation figure prominently in the agenda for the 1992 United Nations Conference on Environment and Development in Rio de Janeiro, Brazil, where conference planners are seeking a substantial increase in financial support for research in biodiversity. It is imperative that biologists and the public learn more about the importance of biodiversity and its role in ecosystem functions.

Only an international collaborative effort that is supported by adequate resources can accomplish this task. Increased public awareness of the serious implications of humanity's depletion of biodiversity before its importance is understood is helping to create a climate that may stimulate governments to support national and international efforts in this field. Biodiversity is the variety of the world's living species, including their genetic diversity and the communities and ecosystems that they form.

Today, this diversity of life is facing serious threats. Scientists concur that, at current rates of tropical forest loss and disturbance, roughly 5 to 10 per cent of all tropical forest species will be lost per decade during the next quarter century. The extinction crisis is not restricted to tropical forests. Freshwater habitats are being altered dramatically as rivers are impounded and exotic species introduced. In the southeastern United States, for instance, 40 to 50 per cent of the freshwater snails are extinct or near extinction.

Oceanic islands, where most of the extinctions of the past several thousand years have occurred, also remain among the most threatened ecosystems on Earth. Some 60 per cent of the plant species native to the Galapagos Islands are threatened with extinction. Although the numbers of species that will be lost are greatest in tropical countries, the proportion of species at risk in the temperate zone is still large.

In 1988, the Centre for Plant Conservation in Jamaica Plain, Massachusetts, estimated that more than 10 per cent of U.S. Plant species and subspecies are in danger of extinction in their wild habitats. The loss of species and genetic diversity and the degradation of habitats and ecosystems represent an enormous cost to this generation and all future generations.

Biological resources—the portion of biodiversity that is of actual or potential use to people— provide the basis for most human enterprises. The species being lost today possess unknown food, medical, and industrial uses. The ecosystems being degraded through mismanagement of biological resources are losing their capacity to support the human populations dependent upon them, and this degradation is exacting further costs through soil erosion, salutation of reservoirs, local climate changes, desertification, and loss of productivity.

Also, the erosion of genetic diversity among agricultural crops through such practices as the substitution of high-response cultivators for local varieties hinders efforts to breed improved varieties. High response vultivars often handicap small farmers who cannot afford the necessary expensive inputs like fertilizer and pesticides. The fundamental social, ethical, cultural, and economic values of biological diversity have long been recognized in religion, art, and literature. The great interest of children in nature, the numerous wildlife clubs, the generous donations made to conservation organizations, the political support for green parties, and the popularity of zoos and wildlife films are personal and economic expressions of preference that show that the public does not think of biological resources merely in terms of their cash value.

Nevertheless, policies that promote conservation can better complete for the attention of government and commercial decision makers if their advocates demonstrate the economic contribution that biological resources make to development. Even partial valuation of the benefits of conserving biological resources can provide, at least, a lower limit to the full range of benefits that conservation contributes to national economic development.

Three main approaches have been used to determine the value of biological resources:

— Consumptive-use valuation involves assessing the value of resources, such as firewood, fodder, and game meat, that are consumed directly, without passing through a market.

— Productive-use valuation involves assessing the value of products that are commercially harvested and marketed, such as timber, fish, game meat, ivory, and medicinal plants.

— Non consumptive-use valuation involves assessing the direct values of ecosystem functions, such as watershed protection, photosynthesis, climate regulation, and soil production, along with the intangible values of keeping options open for the future and the pleasure of knowing that certain species exist.

Such methods of valuation reveal that biological resources confer much larger economic benefits than is typically appreciated. For example, some 84 per cent of the Canadian population participates in wildlife-related recreational activities (a consumptive use), which provide Canadians with benefits that they declare to be worth $800 million (U.S.) annually. In Sarawak, Malaysia, a detailed field study found that wild pigs harvested by hunters had an annual market value of about $100 million. Policymakers often appreciate productive-use values more than consumptive - or non consumptive-use values relate to large industries, such as forestry and fishing, which are connected to markets and, therefore, reflected in national income accounts. Yet few people appreciate the magnitude of the contribution of biological resources to national economies.

One groundbreaking study that showed how to estimate the dollar value of biological resources and analyzed the contribution of wild plant and animal species to the U.S. economy concluded that some 4.5 per cent of the U.S. gross domestic product is attributable to wild species. In the late 1970s, 4.5 per cent of the U.S. gross domestic product was about $87 billion per year.

Although indirect values of biological resources are the most difficult to estimate, they actually may far outweigh direct values. Species without consumptive—or productive-use values may play important roles in the ecosystem and support species that are valued for their productive or consumptive uses. For example, the retention of a wetlands complex outside Boston, Massachusetts, provides an annual savings of $17 million in flood protection alone —excluding the many other benefits, such as sediment reduction, fish and wildlife production, and aesthetic values, that the wetlands afford area residents. In Thailand, the $1.5-million cost of protecting the watershed of the Nam Pong Reservoir more than justified by the lessened rate of reservoir siltation, which requires expensive dredging. Already enormous, the value of biological resources to humans will grow as those resources shrink through environmental degradation.

Diversity in genes, species, and ecosystems provides the raw materials with which human communities adapt to change. Thus, the loss of each additional species, gene, and ecosystem reduces the ability of nature—and people—to adapt to the changing environment.Where does the blame for the loss of biodiversity and the degradation of biological resources lie? On the surface, the answer seems clear. The proximate causes include large-scale clearing and burning of forests, excessive harvesting of plants and animals, indiscriminate use of pesticides, draining and filling of wetlands, destructive fishing practice, air pollution, and the conservation of wildlands to agricultural and urban uses.

The root causes of biodiversity loss are found in basic economic, demographic, and political trends. These root causes include the human demand for commodities, such as tropical hardwoods, wildlife, fiber, and agricultural products; the growth in human population, which, even without proportional economic growth, places increasing demands on natural resources and ecosystem processes that are already impoverished and stressed; settlement policies that promote the migration of the growing unemployed labour forces to frontier zones; debt burdens that force governments to encourage the production of commodities that can earn foreign exchange; energy policies that encourage inefficiency in many nations and, in so doing, add to the burden of air pollutants and the risk of substantial global climate change; and inappropriate land tenure arrangements that discourage rural people from investing in substainable agricultural practices. When biodiversity loss is defined as a result of its immediate causes, the response is to take defensive and, often, confrontational actions, such as enacting laws, closing access to resources, and protecting additional areas.

Such responses are necessary in times of rampant over-exploitation, but they are seldom really sufficient to change the underlying social and economic causes. Defining the problem in terms of its root causes can stimulate a more constructive response that seeks cooperative efforts to address the social and economic foundations of resources depletion. Six main obstacles make it hard to conserve biological diversity:

— National development programmes undervalue biological resources monetarily.

— Over exploitation of biological resources yields great profit for traders and manufacturers (who can externalize environmental costs) while impoverishing the local people who have few other sources of livelihood and who must pay the environmental costs of overexploitation.

— The species and ecosystems upon which human survival depends are still poorly known.

— Scientific research often does not meet the needs of resource and protected-area managers.

— Conservation activities tend to be focused too narrowly.

— Institutions assigned responsibility for conserving biodiversity have lacked sufficient financial and organizational resources to do the job.

These obstacles block progress in both developed and less developed countries. However, because the tropics harbour a major proportion of the planet's biological diversity and because overexploitation of resources in the tropics is linked to resource demand in the industrialized countries, the most pressing needs for action to conserve biodiversity are in developing countries.

Developed countries depend on tropical resources for industrial materials, breeding material, pharmaceuticals, tourism destinations, and numerous other tangible and intangible benefits. However, the exploitation of the tropics by the industrialized countries has yielded great benefits without commensurate investments in conservation and without compensation for the environmental costs of overexploitation. Such factors as cheap labour, artificially low-priced raw materials, inappropriate development aid, and the control of commodity prices and interest rates have encouraged rapid resource depletion and destruction.

The situation is worsening continuously through the ramifications of the developing world's debt crisis and related high interest rates. Conservation will succeed only if both proximate and ultimate causes of environmental degradation and species loss are addressed.

The complex threats to biological diversity call for a wide range of responses from many private and public sectors, and the mix of these responses must be adjusted to suit local conditions. Because government policies often are responsible for depleting biological resources, policy changes usually are a necessary first step toward conservation. National policies that deal directly with wildlands management or forestry or that influence resource use indirectly by affecting, for example, land tenure, rural development, family planning, food prices, pesticide use, or fuel consumption can have significant impacts on the conservation of biodiversity.

National and subnational conservation strategies can often provide the mechanism for identifying necessary changes. The best way to protect species is to *protect habitats*, and most national governments have established legal means to do so. These means include creating national parks and other types of reserves (some 4,500 major reserves exist covering nearly 500 million hectares); enacting local laws to protect particular forests, reefs, and wetlands; restricting private concession agreements: and establishing zoning regulations. The responsibility for managing protected habitats often is spread widely among public and private institutions.

Although accomplishments to date are impressive, the amount of habitat protected must be increased by a factor of three if it is to make the necessary contribution to the conservation of biological diversity. Many of these new areas will need more flexible approaches to management than are customarily used in national parks.

Protected habitats will succeed in realizing their conservation objectives only to the extent that the areas themselves are managed effectively and that the land surrounding them is managed in a way compatible with those objectives. Implementing such management typically involves making protected areas parts of larger, regional schemes designed to conserve natural resources, induce sustainable human use of the resource base, ensure that benefits of the habitat protection reach the rural population, and involve the rural population in policy decisions related to the protected areas. Ex situ preservation programmes, such as zoos, aquaria, seed banks, and botanical gardens, supplement in situ conservation by providing for the long-term storage, analysis, testing, and propagation of a few threatened and rare species of plants and animals and their propagules.

Ex situ preservation is particularly important for wild species whose populations are greatly reduced in numbers and serves as a backup to in situ conservation, as a source of material for reintroductions to the wild, and as a major repository of genetic material for future breeding programmes of domestic species. Some ex ritu facilities — notably zoos and botanical gardens — provide important opportunities for public education, and many make important contributions to taxonomy and field research. Curbing the pollution of the biosphere is the biggest and most expensive conservation effort. Biological diversity is threatened by various forms of chemical pollutions, but ultimately, the gravest threat may be global climate change.

Mean world temperatures could increase by about 2^{0}C in the next 40 years with an attendant rise in sea level of between 30 and 50 centimeters. Although the species and ecosystems within protected areas will certainly be affected by climate change, it is unrealistic to expect the boundaries of existing protected areas to be moved to track the changing climate because these areas usually are surrounded by more intensive human land uses. Instead, new forms of intervention will be required to maintain the ecosystems deemed desirable. Many response to the loss of biological diversity have been supported by international legislation that has fostered global cooperation in conservation, as has the Convention on International Trade in Endangered Species, for example.

However, species and ecosystems are still being exploited at rates that far exceed their sustainable yield. Recognizing the growing severity of threats to biological diversity and the increasingly international nature of the actions required to address the threats, the World Conservation Union(IUCN), based In gland, Switzerland, and the United Nations Environment Programme (UNEP), based in Nairobi, Kenya, have begun to prepare an international convention on the conservation of biological diversity. This effort has gained broad support from many governments. For example, in 1988, the U.S. Congress passed a joint resolution, H.J. Res. 648, in support of the conservation of biological diversity.

Any strategy to foster the sustainable use of biological resources must be based on public involvement. Indeed, initiatives that do not involve local communities generally are doomed to fail. Indigenous people in many parts of the world are especially reliant on natural resources for their cultural continuity and economic well-being. Their role in conservation should be given particular attention, and they should be given opportunities to participate as major players in the design of conservation programmes affecting their resources. Local people should be closely associated with the authorities responsible for both the management of biological resources and the establishment and management of protected areas.

Effective action must be based on accurate information, and the more widely shared the information, the more likely it is that individuals and institutions will agree on the definition of problems and solutions. Developing and using information is, therefore, an essential part of conservation at all scales, from the local to the global community. The current state of knowledge about species and ecosystems is woefully inadequate; detailed knowledge is still lacking on the distribution and population sizes of even such large and well-studied animals as African primates.

It is self-evident that increasing knowledge about the kind and variety of organisms that inhabit the Earth and the ways that these organisms relate to each other and to humans must be a foundation of conservation action. This basic inventory and research should be carried out simultaneously with field conservation efforts; the two activities can be mutually reinforcing. To gain the necessary knowledge, scientists must

— Conduct taxonomix surveys to document the wealth of the world's plant and animal species;

— Undertake greatly expanded ecological research to determine how the various pieces fit together, explain the population dynamics of species of particular concern, assess the effects of fragmentation of natural habitats, and determine what management steps are required to enable ecosystems to flourish with their full complements of species;

— develop and improve mechanisms for ex situ preservation, including both captive propagation of species and their eventual release into natural or restored ecosystems;

— monitor the changes in species diversity and ecosystem function as human influences become more pervasive;

— assess the ecological differences between relatively large but minimally disturbed ecosystems and ecosystems that have been heavily affected by humans to learn how to enhance productivity and restore degraded ecosystems to a more productive state; and

— conduct social science research to determine how local people manage their resources and how changes in resource availability and land use affect human behaviour.

Government agencies, local communities, and conservation organizations all need information to manage biological resources more effectively! Tools that can help meet this need include basic descriptions of flora and fauna, practical handbooks for field identification, rapid inventory techniques, and refined computer programmes for species and habitat inventory. The need for information about the tropics is particularly important because these areas hold most of the world's biological diversity and are losing species faster than scientists can record them.

The highest priority for basic inventory work should be given to the most threatened sites of greatest diversity and local endemism. The information contained in the species in these areas could disappear before humanity even knows what it is losing. Development agencies should support national efforts to establish local, sectorial, and national information management systems by demonstrating methodologies, providing training opportunities for taxonomists and biologists, and subsidizing the publication of status reports. Universities, research institutions, and nongovernmental organizations need to be strengthened so that they can help governments assess their biological resources.

Closer working relationships should be established between museums and other taxonomic-oriented institutions and those concerned with conservation of biological diversity. Is there anything in common between the familiar *neem tree* and a heated global debate over protecting the environment? Or between its overwhelming medicinal and pesticidal attributes and a multi-million dollar industry in a rich and developed nation? Yes, surprisingly. For this hardy tree, taken for granted as part of the Indian landscape, is the focus of scientific research in the USA because of its amazing array of *pesticidal properties*. And as studies exploit the goodness of neem via biotechnology, it is likely that somewhere along the line the indegenous people who discovered the bitter magic of this tree will be forgotten, just as its genes begin to show their might in the dollar avatar. Is this an instance of the sharing of a global resource? The debate over this biological wealth deepens because most of it, on a global scale, originates in the developing, tropical nations of the Third World, which seem to be at cross-purposes with the developed nations who have adopted a big-brother attitude.

The rather verbose draft convention has the laudable objectives of conserving biological diversity by ensuring sustainable resource use, providing funds and transfer of technology-to developing nations for conservation, and the fair sharing of the benefits of research in biotechnology arising from conservation of biodiversity. Lengthy and longwinded, the draft has 40 articles, while the overuse of technical jargon coupled with convoluted sentences makes it almost incomprehensible.

In the absence of a complete understanding of all its implications, India, as one of the world's biologically wealthy nations, is treading softly, even as the minister for environment and forests, Mr. Kamal Nath, is voicing his objections. But what does India stand to gain from the Convention? Will it protect India's biodiversity from the new global order of bio-technological imperialism? India certainly has a lot to protect. "Our country is extremely rich in genetic resources with about 15,000 known species of flowering plants," says Mr.R.S. Rao, Director, National Bureau for Plant Genetic Resources, New Delhi, the country's premier institute devoted to gene conservation and collection.

Imagine, then, the amazing array of candidate genes that must be available for biotechnological manipulation. Says Swarahj who heads the insect research division at the Indian Agricultural Research Institute, New Delhi, "Insects constitute a mind-boggling 80 per cent chunk of all living organisms in India, there being roughly 60,000 different species. Many of these, like the moth which weaves silk, have revolutionized human civilisation.

Then there are the microbes, bacteria, fungi and the multitude of sea life, all vital sources of medicines and life-giving compounds." The list is endless, chock-full of incredible genes from nature, which is why India wishes to understand fully what she stands to achieve from the Biodiversity Convention. Technology transfers and financial assistance of course, say the pundits. It is this link between biotechnology and bioresources that has reached some consensus now, says Mr. Kamal Nath in an exclusive interview soon after his return from the update meeting in New York. It's here that the trouble begins as the patronizing attitude of the West could well mean a 'recolonisation of India'. Points out Ashish Kothari, an observer from the non-governmental sector at the Convention's last meeting in Nairobi, "Technology should be exchanged and not transferred one way since there are a number of appropriate, indigenous technologies which are already underway in India." Of course, there's no doubt that India's level of technological development is strong. Even so Mr.V.L. Chopra, Director General, Indian Council for Agricultural Research, New Delhi, says, "It may be wrong to take a blanket view and reject all foreign assistance with regard to technology.

Instead there should be a case-by-case assessment of the technologies being transferred with a view to retaining the sound and rejecting the useless. Says Mr. Rana "There are two categories of nations—the 'gene-rich, technology-poor' and the 'gene-poor, technology—rich' nations. India falls into neither of these categories being 'gene-rich and technology-rich. Why then the complications of this convention? "The very basis of biological wealth," says Mr.Chopra, "has been the general concept that the free flow of germplasm on an international scale should continue." But who will call the shots? Chhatrapati Singh of the Indian Law Institute, New Delhi, an expert on environment law, feels that the convention is yet another example of Western formulation of international agreements which, far from being globally participatory, are mere 'sign-on-the-dotted-line' documents.

Also, the sharing of biological wealth must be judged in totality along with the issues of intellectual property rights, the GATT and patents. Even Mr Kamal Nath believes in the need to protect our biological wealth via IPRs and patents. Here one comes full circle to the neem tree. India is a treasure-house of living things and has a wealth of genetic resources in its umpteen biographical zones—the deserts, coastal areas, rain forests, teak and sal forests and the meadowlands of the mountains.

Does their conservation really require a Convention? Yes, say spokespersons of the ministry of environment and forests, but not a convention which will cost the nation its sovereignty. There is also the feeling that the Convention will focus more on the utilisation of resources than their actual conservation. Can we then set our own terms without foregoing our independence? Says Mr.Rana, "One way to check the imperialism being forced on biologically wealthy nations like India is to strengthen our gene banks and research base, inventorise and characterize our genetic wealth and invest it in basic research."

Realising that there are several shortfalls in government programmes, M.S. Swaminathan, President, World Wide Fund for Nature (WWF), launched a people's biodiversity programme. Under the aegis of the Jawaharlal Nehru Community Biodiversity Conservation Movement at the WWF, these projects have managed to involve effectively the local communities in reforestation, nursery work and wildlife conservation programmes all over India.

Several such programmes have been highly successful. Some even suspect the motivation of the unusually generous offers made by certain developed nations with regard to financial technological assistance. Even as Mr.Kamal Nath declares India's decision to reject the Convention unless suggested changes are made, India might continue to attend the negotiations and pre-Rio meetings. Will India be lured by these attractive offers of monetary help? Or will she continue to guard her national interests and biodiversity from the developed nations' hegemony? Though Mr Kamal Nath protests that India is not on dole, only time will tell how these questions will be answered ultimately.

Meanwhile, the draft convention high-handedly proclaims awards to the originators of traditional methods or using biological wealth. In reality, there are so many illegal outlets for both indigenous genes and their utilisation methods that losses occur every day. So, if neem is today the catalyst of a dollar machine, it may have been another plant yesterday and might be yet another tomorrow.

A faulty conservation strategy with one-sided formulations may only further the damage, making poorer nations the 'beneficiaries' of low-grade, obsolete technologies. These losses may far outweigh the gains from such a Convention.Yet there are the silver linings. Mr. Kothari feels that with negotiation and substantial modification, the Convention could be a useful tool for the conservation of biodiversity since it would provide a vital link between nations, with proposed global lists avoiding duplication while ensuring the conservation of all species. "The balanced discussions between the 80 participating countries have been a definite step forward in the chequered history of environmental movements," he points out. "Further, we should adopt a holistic view of the Convention without rejecting it outright.

After all, one never knows when India might be in dire need of a gene not to be found here." Though the facts are clear, the implications are not. There are too many spokes in the wheel. One worrying aspect is that a great majority of people—even people in key positions—seems to be ignorant of the convention's text. Its political understones are strong and could prove overwhelming.

The fault is perhaps in the preliminary formulation of the draft carried out at the World Resources Institute, Washington, DC. If participatory drafting involving all members—'contracting parties' as the Convention calls them—had been done, some of the hurdles could have been avoided. And yet, Mr. Kamal Nath is of the opinion that with the G-77 group of developing countries backing India's stand, there is room for optimism. Meanwhile, the 'fair and equitable sharing of resources' seems to have become the golden phrase now. So Utopian does this seem that one would imagine it will never materialise.

Earlier instances of gene hunting in the tropics by Western nations are far from being rosy stories of North-South cooperation. Perhaps, the goodness of neem needs protection in more ways than one. Although there are a number of agroecosystem management approaches which provide alternatives to these trends, Bio-Dynamic Agriculture, Fukuoka's Natural Farming, organic agriculture 'alternative' agriculture sustainable' agriculture 'ecological' agriculture Holistic Resource Management is of particular interest in the context of biodiversity issues because of its emphasis on biodiversity protection and enhancement as a central management principle. In Holistic Resource Management courses, students are encouraged to work with fundamental ecosystem processes and increase biodiversity on their farms or ranches to enhance profitable sustainability.

Savory builds his argument for greater biodiversity in agroecosystems on the debated hypothesis in ecology that stability is positively correlated with diversity in ecosystem function. McNauhton's work in old fields, East African grasslands, and more recent work by Tilman and Moffat support Savory's contention. The fundamental ecological premise of HRM is that efficient nutrient cycling, hydrology, successional dynamics, and energy flow, with a strong biodiversity base both above and below ground, buffers managed ecosystems' responses to perturbations.

Theoretically, by learning to work with and improve biodiversity and these ecological processes on their land, farmers and ranchers need less external inputs to sustain production, which then allows them to reduce their input costs, enhance their profitability and sustain their resource base for the long-term future. It is our idea that biodiversity, used as a management concept, could impact ecological processes at population, community and ecosystem levels.

For example, while diversified cropping practices, in both space and time, can contribute to increased populations of natural enemies, the same practices offer opportunities to recycle nutrients within a farming system and, also, present opportunities to value add enterprises within a farm. Perhaps most fundamentally, a decision process such as HRM provides a framework that encompasses a diversity of practices, a ranee of options that a farmer can then 1 best adapt to local environmental, economic and social conditions. For these reasons, we believe it is important to view biodiversity as a multi-function management tool.

We desired to learn how exposure to HRM influences attitudes toward biodiversity among farmers and whether the actual experiences of HRM practitioners support Savory's idea that biodiversity is the foundation of profitable sustainability in agriculture.

We used a participatory ethnographic approach and qualitative research methods to explore the attitudes, perceptions, decision making, observations and experiences of a group of farmers and ranchers who had proven expertise in using HRM. Our data are based on the practical experience of these land mangers rather than experimental scientific data. Cheekland and Bawden et al. describe two different approaches to agricultural research.

The *scientia* approach is quantitative, focused, precise, slow, expensive and, if done in optimal conditions, so controlled as to be disassociated from the complexities of reality. It focuses on mechanisms and predictability. In contrast, the praxis approach is qualitative, rich in contextual information, relatively inexpensive, quick and highly confounded with complexities of the real world. While many agricultural scientists are accustomed to scientia information, a growing number are recognizing the value of the kind of information produced in praxis studies, which can provide critical insights into the context of a particular area of scientific inquiry.

Although biodiversity may be viewed as a fundamental principle of sustainablity by many scientists and environmental activists, in our experience with farmers in Ohio, Missouri, Kansas.

West Virginia and Vermont, at least, it is rarely taken into consideration. If we wish to develop more forms of agriculture which protect biodiversity, we need to understand the perspectives of the people who manage agricultural lands in general. However, in particular, understanding the perceptions and experiences of land managers who are committed to these principles can be a valuable step towards developing research and education for a wider audience.

Other researchers working in a participatory mode with farmers throughout the world have found that indigenous experience and knowledge have provided an effective foundation for research and development of more sustainable agricultural practices. The Holistic Resource Management process is being used by farmers, ranchers, communities, government agencies and, increasingly, businesses, nationally and internationally. Holistic Resource Management evolved out of Savory's concern for the alarming rate of desertification he observed as a wildlife biologist in his native Zimbabwe and many other parts of the world.

Savory equates desertification with loss of biodiversity. HRM students are taught that biodiversity is the source of prosperity and that loss of biodiversity can have far-ranging environmental and social effects. 'Biodiversity is not about rare and endangered species, it is about human survival'. Savory argues that the fundamental cause of biodiversity loss is the process of how management decisions are made and he offers the Holistic Resource Management Model as an alternative decision making process. The first step in the HRM process is to define thee 1 whole' being managed in terms of people, landbase and money.

Once a clear assessment has been made in these terms, the people directly involved should work together to develop a holistic goal. This goal includes: (1) their quality of life values, (2) forms of production they must achieve from the land or other resource base to support each of their quality of life values, and (3) a vision of what they wish the land or resource base to look like in the future to sustain their production and that of future generations.

In the future landscape description, HRM students are taught that it is important to understand and work with four ecological processes as foundation blocks community dynamics, the water cycle, mineral or nutrient cycles, and energy flow. Only after this contextual groundwork has been laid do specific tools come into the HRM process. Although most people, and especially natural scientists, are much more comfortable discussing tools than quality of life values, it is tools that we most often argue about. Laying the context of a farm or ranch first can be a key to building, unity in otherwise disparate groups of people in the whole.

The tools used in the HRM model include: human creativity, money and labour, rest, fire, razing, animal impact, living organisms, and technology. Creative brainstorming or idea generation is encouraged to produce a list of possible solutions to management challenges. These, then, are tested using seven testing guidelines in a pass or fail mode, against the established goal.

— The first testing guideline, Whole Ecosystem, asks how the proposed tool or enterprise will affect the four ecosystem processes; will they be enhanced or degraded and will it move these processes towards or away from the future landscape description?

— In the second testing guideline, Weak Link, the user determines the weakest link in their operation. Is there a lack of knowledge or information that is holding back progress toward the goal? Is the weakest link in personnel or financial problems, inadequate or poor land? Where is the weakest link in the energy conversion chain from sun to plants, to products that can be marketed? Is there a problem organism involved? If so, where is the weakest link in the life cycle of the organism?

— In the third testing guideline, Cause and Effect, the user considers whether the proposed action will treat a cause or symptom.

— The fourth testing guideline, Marginal Reaction, is one of the economic guidelines and is used only when comparing more than one option. It asks which option will provide the biggest return for the money or time invested in moving the operation towards the goal.

— The fifth testing guideline, Energy/Wealth Source and Use asks: Will the proposed tool require the use of finite sources of energy and will such use have to be repeated? The fifth testing guideline asks also: What is the source of money required paper dollars, mineral dollars (money from non renewable resources) or solar dollars.

— The sixth guideline, Society and Culture, asks how the proposed tool or action will affect the culture and society. Will it help strengthen the community?

— The seventh and final testing guideline, Gross Margin Analysis, is another economic test and, like Marginal Reaction, compares several options., Using the approach of British economists, Wallace and Burr, in which only variable costs are figured, this guideline helps determine which forms of production are most profitable.

If used conscientiously, this decision making process gives the decision maker a consistent and objective evaluation of the soundness of the decision and helps insure that any decision is optimized between environmental, economic and social considerations. Savory argues that it is this conscious optimization process that sets HRM decision making apart from decision making that humans have been using throughout our history as a species and regardless of culture.

The Management Guidelines part of the model deals With specific and more advanced aspects of Holistic Resource Management. See Savory and Bingham and Savory for additional information. In the last part of the model Control, Test Assume Wrong, Early Warning Criteria, Monitor, etc., the user assumes the decision is wrong and watches for the earliest sign of deviation from the plan, be this economically, ecologically or with respect to quality of life. This, approach creates a different psychology from that in conventional decision making in which a carefully researched decision generally is assumed to be right.

The names and contacts of committed Holistic Research Management practitioners were requested from established certified HRM educators and the Center for Holistic Resource Management in Albuquerque, New Mexico. We wanted to interview a range of HRM practitioners with respect to several parameters: (i) length of time using Holistic Management, (ii) different climatic regions or position on the 'brittleness scale', (iii) scale of operation, and (iv) enterprises. A set of questions was agreed upon. Potential interviewees were contacted by correspondence and given the opportunity to think about the questions before being called and interviewed. A few of the interviewees chose to respond in written form and four individuals were interviewed in person, but the majority were interviewed by telephone.

The interviews took an average of one hour to conduct. Biological monitoring is the part of Holistic Resource Management which detects early warning signs of ecological problems. Savory and Bingharn and Savory present methods in which attributes concerning plant cover, density and species composition and soil surface conditions are measured on fixed transects. Instead of direct statistical analysis, the data are related to the HRM model for interpretation. In addition to providing information on changes in plant community composition over time, this monitoring offers the manager a relatively simple and inexpensive means (in comparison with scientia research methods) of assessing feedback on parameters associated with all the ecosystem foundation blocks in the model. Random sampling points are determined along fixed starting point transects using a dart thrown over the shoulder. Sampling is to be done at the same time of the year each year by the same person. Less than 5% of the people interviewed were doing the full scale biological monitoring recommended by Savory and co-workers.

Most of them relied on less time-consuming methods of monitoring their land ecologically, which included walking their land and getting down to ground level and observing soil and plants. Observations of faster decomposition of manure piles, more earthworms and other soil macro-fauna and changes in organic matter or soil tilth, for example, were viewed as indications of changes in nutrient cycles. Many interviewees reported observing changes in soil erosion and hydrological cycles, such as springs and streams running where none bad run for many years and clear instead of murky water in ponds. Another indication used for changes in the water cycle was both resistance and resilience to drought. Finally, for changes in energy flow, we used reported increases in plant biomass and increase in stocking rates of livestock.

Stocking rates of livestock on a unit of land was a parameter reported by numerous interviewees and was viewed as a dirèct indicator of profitability in addition to an indirect indicator of energy flow. Stocking rates are the amount of land required to support one animal for a specified period of time, usually a year. The scale of operations ranged from 7.3 ha to 90000 ha and the time practising HRM 1.517 years with an average of 5.6 years.

The cases were characterized by wide range of farm/ranch types. All but one of the cases were using planned intensive grazing on some of their land, if not all. Interviewees typically defined biodiversity as all the different species of plants and animals, including microorganisms in the soil, and their quantities as biomass or numbers. In addition, habitat diversity and genetic diversity within agricultural species were often mentioned. One interviewee included diversity of human ideas in his definition of biodiversity. Nine percent of the interviewees thought about biodiversity in the context of their operations before being exposed to HRM. Now, 100% of them think biodiversity is important to the ecological and economic wellbeing, of their farms or ranches. Biodiversity is the true wealth to sustain the world. It is the only true source of capital for enterprises.

We were not aware of biodiversity before HRM, said one of the interviewees in response to question 6. All but one interviewee reported observing increases in biodiversity since they began using HRM. With respect to changes in profit, 80% reported increases from their land since HRM began influencing their decisions. Of these, 40% provided actual percentage increases which ranged from 60% to 1400% and averaged 386%. Sixteen percent reported no or little increase in profits yet, but were optimistic about future increased profitability. In addition to reported increased biodiversity on their farms and ranches and concomitant reported increases in profitability, almost all of the interviewees reported observing improvements in ecosystem processes since they bean using HRM, For example, a farmer/rancher in North Dakcta with 10 years' experience practising HRM, related increases in soil permeability and infiltration from 5 em to 50 em.

All of the ranchers west of the Mississippi River reported changes in plant species composition to a greater frequency of perennials and the return of many native tall and short grass prairie species. On the human community side, most of the interviewees relayed that they were considered odd by their immediate neighbours, but many of them are sought after speakers for national meetings and even internationally. Networks with other HRM practitioners were considered critical to the success of most of the interviewees. With respect to quality of life issues, 91% of the interviewees reported that HRM has influenced some aspect of time management in their operations and lives in general, of this group, 29% reported no change in total labour, but that how that time is allocated has changed, 'I don't know that the amount of time we spend has changed, but how we spend it has', said one of the interviewees. 'We are no longer doing a lot of things we do not like to do', reported another. Fifty-two percent, reported decreases of up to 40%-60% in labour requirements in their operations, in spite of the extra planning and monitoring required by HRM. 'Not only do I have time to go out to eat with my family one night a week now, but I can pick which night', one of the ranchers told a group of HRM workshop participants. A

All of the ranchers interviewed on the original tall and short grass prairies of the USA reported the return of native species in their intensively managed grazing cells and reported that their cattle were doing well on the native forage. Ecologically, this is very interesting, in that it was agriculture that caused degradation of the American prairies and on these HRM managed lands at least, agriculture is helping to restore the native ecological communities.

These ranchers believe that their investments of time and money to upgrade their land is netting increased profits in higher carrying capacity and lower production costs. These three operations offer a range along Savory's brittleness scale, with the farm in West Virginia in the humid southeastern part of the USA being in the least brittle environment, followed by the Oklahoma Coffey Ranch, and the New Mexico ranch in the and southwestern part of the USA in the most brittle environment.

The Fichtner family moved to this 32 ha hillside farm in 1981. At that time, Fichtner reported that the farm was almost overrun with multiflora rose and its soils were classified as severely eroded and very severely eroded. It had been plowed for maize production in the 1930s and then put into sod in the 1940s or 1950s and the land supported a few horses and beef cattle since the 1950s. The farm is located in an area which was fire-maintained savanna grazed by the Eastern Woodland Bison before European settlement. The Fichtners began intensive grazing in 1989 and discovered HIM at an intensive grazing conference The husband and wife took their first HRM classes in 1990 and 1991, respectively. Both had off-farm jobs at that time. With their goal and the principles and tools of HRM they began to reclaim their land. A great diversity of animals are used as tools each with their own role. For example, the hogs break up and compost manure in the barn and the Muscovy ducks control flies. The Scottish Highlander cattle are very rugged and excellent browsers clearing brush efficiently. The donkeys keep coyote at bay.

The cattle graze after the sheep, helping to break the parasitic cycle, as well a harvest the plants their predecessors passed over Lambs are pasture born in the spring. The Fichtner began controlled grazing in 1990 on a small seal with seven paddocks, a few sheep and electric net on a 0.2 ha field.

2

Biodiversity of Social Biota

Soil macrofauna biodiversity was studied during the May-October periods of the years 1978-1991 in the following set of ecosystems located in the territory of the Central Chernozem Reserve and the Kursk Agriculture Station: 120 ha of natural unmown meadow steppe; 500 ha of mown meadow steppe; 60 ha of pasture 45 ha of barley field without fertilisation 1 ha of permanent black fallow ploughed twice every year. The mowing cycle in the mown steppe is hay-making once per year for three consecutive years followed by one year without hay-making.

The meadow steppe and pasture ecosystems are the only grassland ecosystems in the Central Russian region that have never been ploughed during human history. They were used as horse pasture or hay-making land until the 1930s, when they became part of a reserve. Other small islands of real steppe can be found in Ukraine. These ecosystems were thoroughly investigated by climatologists, soil scientists, botanists, geochemists, vertebrate zoologists and others for long periods of time, and a lot of data have been published on the chemical composition and long-term dynamics of the ecosystems.

Plant biodiversity studies have been carried out for more than 60 years while soil animal biodiversity research began about 40 years ago. The main studies on soil chemistry were carried out in 1950 1980, plant chemistry in 1950-1980 and animal chemistry in 1970-1980 in the same ecosystems. In most of our studies we used the following method of soil macrofauna estimation. Soil samples were taken by soil corer up to 30 cm depth.

Every 10-cm layer core was put into a plastic bag, brought to laboratory and fauna was extracted by hand sorting. After the extraction, animals were kept on a moist filter paper for 24 h to excrete the gut content. They were then dried at 100°C in a thermobox for 48 h and weighed. Methods of chemical analysis of the soil, plants and animals collected have been described in detail elsewhere. Earlier we had collected soil samples of 50 cm × 50 cm × 30 cm and of 25 cm × 25 cm × 30 cm and extracted animals by hand sorting in the field.

However, it was found that collection of these large samples led to disturbance of the vegetation cover around every sample of an area four to six times larger than the area of the sample. This sampling procedure could change the vegetation cover in and around places of collection and alter natural processes in the unique reserve of ecosystems (we collected 10 samples every 2 weeks from May to October in the unmown steppe and a reserve oak forest). It was therefore decided to modify the sampling procedure, i.e. sample size and number of samples.

Twenty samples of 10 cm × 10 cm every 2 weeks were considered sufficient for the study of seasonal dynamics of soil macrofauna in the reserve habitats. Population numbers of different species of soil macrofauna obtained by this sampling procedure were practically equal to those found with the earlier method. The total number of soil samples collected during the period of investigation was near 3000. This includes all samples gathered during studies cited below.

The species number and population characters of soil macrofauna decreased in the following order: unwown steppe > mown steppe > pasture > barley field > permanent black fallow. It is noteworthy that in the black fallow ecosystem only one or two species from any of the macrofauna groups were found, namely the earthworm Nicodrilus caligonosus, the diplopods Brachyiulus jawlowskii and Megaphyllum rossicum, the staphylinid beetle Amischa analis, and the carabid beetles Bembidion lampros and Microlestes minutulus. None of the geophilids or lithobiids that are common in the other ecosystems were found in the black fallow ecosystem.

The elaterid larvae found in the soils studied were phytovorous or carnivorous, and it is therefore evident that they were absent in the black fallow soil. Only ubiquist species were present in the barley field (the same species of earthworms, diplopods and carabids as in the black fallow soil, as well as the staphilinid beetles *Scopaeus minutus* and *Amischa analis*, the elaterid larvae *Agriotes sputator*, and the lithobiid *Monotarsobius curtipes)*, and super-dominance is the particular feature of the populations in the barley field and black fallow soils. The exception are geophilids species: *Geophylus proximus* comprises 67% of all geophilid specimens, and *Arcto-geophilus macrocephalus* and *Schendula nomorensis* 16% each.

Practically all animals found in the field and the fallow soil are microbivorous or carnivorous. Although grazing pressure at the pasture is relatively low, and along the pasture borders there are patches of steppe vegetation, soil macrofauna biodiversity is low in the pasture compared with the steppe ecosystems., Only the carabids and elaterids show an increased species number in the pasture soil compared with the barley field and black fallow. Amara species of carabids and wireworms are new members of the groups under study in the pasture soil, in addition to the species already mentioned for the barley field and black fallow.

Staphylinid beetles found in the pasture soil were all coprofillous. Regarding geophilids, *Pachyinerium ferrugineum* replaces *G. proximus* and S. *nemorensis* found in the barley field soil. Density and biomass of the soil macrofauna population in the pasture soil is similar to that in the soil of mown steppe, but biodiversity is lower.

Biodiversity is notably greater in the steppe ecosystems compared with the other ecosystems studied and is comparable in the mown and unmown steppe plots. However, density and biomass are markedly higher in the unknown steppe than in the mown steppe. There are no differences in the species composition of earthworms, diplopods, geophilids, lithobiids and elaterid larvae identified in soils of the mown and unmown steppe. Some new species appear in these groups in addition to those mentioned above.

The differences are marked only for population density and biomass of separate species in these groups. E. *nordenskioldi* is the dominant species in the unmown steppe soil and N. *roseus* in the mown steppe soil. On the other hand, the staphylinids and carabids differ markedly in these ecosystems. The Soerensen index for carabids is 0.64 and for staphylinids is 0.52. The Jaccard index is 0.47 and 0.36, respectively. All other community parameters, such as mean species number per sample, mean density and biomass of animal populations, are correlated with the changes of biodiversity. Mean species number per sample shows a stronger correlation with the population density and biomass than does total species number. This is because mean species number per sample is a characteristic of dominant and common species in determined density and biomass.

A decrease of biodiversity of soil mesofauna groups is a common phenomenon in sets of ecosystems with increasing human influence. Among the many possible reasons, soil perturbation, and microclimatic and soil chemistry changes are considered the leading causes Even a low-impact human activity such as hay-making is reflected in the soil macrofauna composition, and differences between the mown and unmown steppe communities are considerable.

Pitfall data collected in the same period indicate that in the mown steppe the share of ubiquists among staphilinid beetles and carabid beetles is 17% and 5%, respectively, in the unmown steppe it is 28% and 7%, respectively, in the pasture 32% and 44%, respectively, and in the barley field and black fallow ecosystems 100% for both groups. It is interesting that the share of ubiquists in the field increased from 72% in the 1960s to 100%. This may be a result of an increase in the human impact on the local environment as a whole. In our opinion the higher proportion of ubiquist species of beetles in the unmown steppe is related to the microclimatic conditions in the thick layer of, litter, where conditions are optimal for ubiquist species.

On the other hand, *E. nordenskioldi is* the only non-ubiquist species among earthworms in both ecosystems but is dominant in the unmown steppe. Two other species are found in all anthropogenic habitats and may be considered as ubiquists, with *N.*

roseus dominating in the mown steppe. Based on data on biodiversity, it is difficult to estimate what type of a reserve regime is closer to the initial prehistoric conditions of real steppe (mown or unmown steppe). The grass litter layer is very thick in the =mown steppe but practically absent in the mown steppe.

Geobotanists consider pasture regimes to be the initial one for real steppe, and the mown steppe regime as analogous to the pasture (grazing) regime. However it is evident then that pasture and mown steppe cannot be considered as analogues. The biodiversity and community characters prove this assumption. Grazing activity in pasture at the level of one cow per hectare is sufficiently high to disturb steppe ecosystems and change them entirely. It is very significant that this level of pasture activity does lead to the disappearance of the grass litter layer.

Many pasture soil species are ubiquists or coprofillous. Our observations showed that such coprophagous beetles as Aphodius and Onthophag*us* species are common in surface soil of pasture. These species are practically absent in the mown and unmown steppe. The point of view that pastures and mown steppe are analogous ecosystems was universally adopted by geobotanists. However, a study of species composition at a steppe plot after artificial fire shows that during the first 2 years soil fauna species composition is similar to that on mown steppe plots.

Then species composition is changing toward that typical of unmown steppe. It confirms our opinion that steppe ecosystems are cyclicpyrogenic ecosystems, and not pasture ecosystems. Therefore the species composition of soil macrofauna in the mown and unmown steppe denotes that these ecosystems can be regarded as separate stadia of post-fire succession series in meadow steppe, and not as an initial state of real steppe ecosystems. Earlier we showed that changes in biomass and production of soil animal populations in agroecosystems depend on available organic matter and nutrients in arable soils Correlation between species composition and population density and biomass suggests that resources of nutrients also have an influence on biodiversity.

The differences in composition and biomass of the soil fauna in arable plots are small although the quantity of organic resources in the field soil are somewhat higher than in the fallow soil In the fallow soil there are no such sources of available carbon, nitrogen and phosphorus as plant roots for animals and microorganisms. Available organic matter for animals in field and fallow soils is mainly present in the form of decomposable humic compounds and microorganisms.

The low biodiversity in field and fallow soils are probably a reflection of small pools of available organic matter, nitrogen and phosphorus sources for the fauna. Stores of available organic matter in the natural steppe are three times higher than in the field and five times higher than in the fallow. The microgen pool in the steppe is 1.5 times higher than in the field and fallow. The phosphorus pools in the different ecosystems are practically equal.

However microbial biomass in the soil and in the tits of soil animals are the main sources of essential nutrients for animals Biomass of microorganisms in field and fallow soils is higher than in meadow steppe soil, but it is only 'potential' biomass. To be the nutrient pool for animals, in microorganisms need stores of available energy in soils. These stores are concentrated in dead plant residues and plant exudates.

Soil microorganisms in the field and fallow are active if there is an input of available carbon compounds. Cellulolytic activity of microorganisms in soil of the fallow was higher than the activity of both soil fauna and microorganisms in soil of the meadow steppe when dead roots or hay were put into the soil. Cellulose for fallow microorganisms was a good source of available energy, while steppe microorganisms probably used more easily available sources of energy.

Hence the biodiversity of soil fauna is possibly correlated with amounts of available organic matter and nutrients, and not only with habitat conditions. In arable soil it will be dependent on input of available carbon compounds for soil microorganisms. If we compare overall biodiversity of ecosystems we usually only consider biodiversity of multicellular organisms and not the biodiversity of microorganisms. Unfortunately, there are some problems with the comparison of biodiversity of taxa of different size levels, and of ecosystems as a whole. Over 10 years ago, we first suggested a new index for biodiversity comparisons.

Our aim was to find indicators for a more objective estimation of biodiversity. In our opinion, DNA in ecosystems could be an indicator of 'potential' biodiversity in an ecosystem as a whole. We called the suggested character the 'index of genetic saturation' (ICS) IGS is equal to the sum S(mjx kj), where S is the sum of the products of mj and *kj* for species of a separate group (S is equal to the species number in the group) in an ecosystem.

3

Microbial Biodiversity

The microbial ecology and diversity of the rhizosphere (root system) has received a great deal of attention, the phyllosphere has been described as an ecologically neglected milieu. Some more attention has been paid to the microbial ecology of the aerosphere. Chen et al. investigated the fungal flora on twigs and fruits of chestnut in many areas of China. Results showed that many organs of chestnut trees were associated with microorganisms. Among the fungi, the most common genera were *Phoma, Dothiorella, Sphaeropsis, Gloeosporium, Cytospora, Alternaria, Rhizctonia, Coniothyrium, Phomopsis and Fusarium,* although the type and abundance of the fungi differed on different organs in different areas.

Fungal diversity decreased in the following order: Southwest China, the Yangtze River catchment, the Huaihe River catchment and Yanshan Mountain area. The study also revealed that the ratio of parasites to saprophytes changed as the chestnut fruits developed. Saprophytes and weak parasites dominated at the early stages of growth whereas the population of parasites increased significantly as the fruits matured. Endophytic microorganisms are defined as all living microorganisms (saprophyte, parasite and symbiont) within the, plant body. Numerous microorganisms occupy cortical pores calyx tube, and core and vascular system. For instance in studies of the microflora of vascular systems in cotton plants, Lu found that the main fungi were five *Fusarium* species, three *Chaetomium* species, *Trichothecium* sp., *Trichoderma* sp., *Alternaria* sp., two non-spore forming species and some yeasts. The predominant bacteria were *Bacillus*, *Erwinia*, *Xanthomonas*, *Corynbacterium* and *Chromobacterium*.

Besides the species diversity of microorganisms (viruses, eubacteria, archaebacteria and eukaryotes), we can distinguish various forms of functional diversity of microorganisms, such as metabolic diversity (aerobic, anaerobic or facultative and phototrophic, chemotrophic, autotrophic or heterotrophic) and diversity of interspecific relationships (commensalism, synergism, mutualism, amensalism, competition or parasitism).

In the field of agriculture, the following functional characteristics of microorganisms deserve particular attention: nutrient supply and growth promotion for the plant ("biofertilization"), antagonism, competition, and mycoparasitism relative to plant pathogens ("biocontrol"), and degradation of pollutants in the environment (biodegradation").Among the nutrient suppliers, nitrogen-fixing bacteria are probably the most important group. They may be free-living (e.g., *Klebsiella pneumoniae)* or associated with the roots of either leguminous plants (e.g., *Rhizobium meliloti)* or monocots *Azospirillum brazilense)*. It has been shown that the effect of nitrogen fixation is most significant for leguminous plants whereas the benefits to non-legumes may be small.

Indeed, the beneficial effect of *Azospirillum* may derive not from its nitrogen fixing properties, but from the stimulating effect on root development. Several bacteria may enhance mineral uptake by the plant by solubilizing soil phosphate. In addition to nitrogen-fixers, also these bacteria may have considerable potential as bio-fertilizers. Many microorganisms are known to produce plant growth regulators (PGRs), including auxins, gibberellins, cytokinins and ethylene. Many microorganisms produce antimicrobial compounds which may act on plant pathogens by, inducing fungistatic, inhibition of germination, lysis of fungal mycelia, or by exerting fungicidal effects.

For example, fluorescent pseudomonads produce siderophores which have both fungistatic and bacteriostatic effects by depriving the pathogen of iron, a limiting essential nutrient. Chitinase produced by some bacteria lyses the hyphal walls of a Fusarium species. Volatile compounds, such as ammonia and HCN produced in the rhizosphere are also harmful to certain pathogens. Phytopathogenic bacteria generally occupy a particular ecological niche, but several beneficial bacteria compete with them for habitat and nutrients. For instance, Pseudomonads and *Bacillus* spp. catabolize diverse nutrients in the root zone or on leaves and have short generation times. They are therefore logical candidates for biological control of phytopathogens.

Some fungi can parasitize on other living microorganisms for nutrients. Perhaps the best-known mycoparasite is the fungus *Trichoderma*, hyphae of which may penetrate resting structures or parasitize hyphae of other microorganisms. This characteristic is of interest for biocontrol of plaint diseases. Because of the intense activity and growth of microorganisms, they have considerable potential to enhance the biodegradation of organic contaminants *present in* the environment, such as pesticides, herbicides, raw fuel oil and industrial pollutants.

Microorganisms could be used for degrading contaminants both in and on plants and in the larger environment. The crop plant and the microorganisms living on and in it constitute a holistic system with a complex equilibrium based on the interactions among the various microorganisms and between microorganisms and crop plant.

Changes in this equilibrium, particularly alterations in the interaction between pathogens and other microorganisms, may affect plant disease occurrence and plant growth. For instance Kloepper reported that most plant growth-promoting rhizobacteria (PGPR) strains that have been investigated to date appear to enhance plant growth indirectly by reducing deleterious rhizobacteria (DRB) populations.

The relationships between plant and microorganisms are the result of coevolution: Host plant evolution is affected by pathogen stress, while pathogen evolution is influenced by host plants. The final result is a stable balance, resistance to crop disease and pathogenicity of plant pathogens tend towards moderation. In the original habitat of the crop plant, plant and microorganisms (including pathogens) lived together for lone periods of time and as a consequence the plant-microorganism system is very complicated and steady. Because of the diversity of microbial species arid function, microbial products are widely used in agricultural production, especially for plant disease control. The use of microbial products has the following advantages over conventional chemicals:

— They are considered safer than many of the chemicals now in use.

— They do not accumulate in the food chain.

— Self-replication of microbes circumvents the need for repeated application.

— Target organisms seldom develop resistance as is the case when chemical agents are used.

— Properly-developed biocontrol agents are not considered harmful to ecological processes and the environment.

At world level, a lot of microbial products have been developed for agricultural purposes and are being rapidly commercialised. For instance, the first US product consisting of a PGPR strain. QUANTUM *(Bacillus subtilus),* is readily available at the commercial level. The concept of microecotechnology, which covers biological control of plant diseases and pests, probiotics of animals and human, and bioremediation of environmental contaminants, has been well-received by many scientists in China. Indeed, microbial products are being extensively studied, continuously developed, and widely applied in China. Microbial products are generally divided into two categories: pure culture microorganisms, and organic soil amendments or conditioners with living microorganisms.

In this section, two cases of the use of microbial products in Chinese agriculture are discussed with an emphasis on their role in preserving microbial balance, promotion of crop growth, suppression of diseases and removal of contaminants. Research on applying live microorganisms to crop production is rapidly developing in China. Most of the research involves the utilization of *Pseudomonas* spp. and Bacillus and undoubtedly one of the most successful examples is yield-increasing bacteria (YIB), YIB is a commercial name for a microbial agent made up of different strains of *Bacillus spp.*, mainly B. *cerues*, and *B. subtilis*, as well as *B. firmus*, *8. lichenformis*, *B. coagulans*, *B. brevis*, and *B. sphaericits.* A group of Chinese phytopathologists led by Yanxi Chen from Beijing, Agricultural University had long studied many plant, diseases, such as black rot disease *(Ceratocystis fimbriatia)* of sweet potato and wilt diseases (*Fusarium*) of cotton. On the basis of their experience, they introduced a concept called the natural ecosystem of the plant, which expressed the idea that an individual plant constitutes a microbial ecosystem, and developed a microbial agent called "yield-increasing bacteria" to improve the conditions of plant growth.

The beneficial effects of YIB were first demonstrated in 1986, According to the Ministry of Agriculture, in 1991 YIB was applied to more than 50 types of crops covering 52 million ha of cropland in 30 provinces and municipalities in China. It was reported that the average yield of crops had increased by 10% or even more following the application of YIB to the field crops. The screening process for YIB has been based on the following considerations: (1) the strain should be capable of adaptation and colonization in the target plant; (2) the strain should be strongly competitive with the indigenous microorganisms; (3) the strain should be beneficial to plant health and growth.

The screening process for YIB consisted of the following steps. First, strains were isolated from the surface or from within the tissues of the plants grown in the field. Then, YIB strains were selected by bioassay tests conducted in greenhouses rather than by traditional screening methods (Petri dish antagonistic test). The parameters for selection included percentage of emergence, final stand, root and shoot fresh weight, number of tillers, and disease resistance or suppression. Finally, strains were tested by replicated field trials.

The microbial agent (YIB) is applied by either dressing seeds, spraying leaves of crops or soaking roots of crops in spore suspension. The application is generally more effective at the early stages of plant growth. Beneficial effects on the crop include earlier seed germination, larger root system, and greener leaves. The application of YIB reduces the severity of some diseases, such as sheath blight on rice caused by *Rhizoctonia solani,* take-all on wheat caused by *Gaeumannomyces graminis* var. *tritici,* wheat scab caused by *Gibberella zeal,* sharp eye spot on wheat caused by *Rhizoctonia cerealis,* black rot on sweet potato caused by *Ceratocystis fimbriata* downy mildew on cucumber caused by *Pseudoperonospora cubensis,* bacterial wilt of sweet potato caused by *Pseudomonas solanacearum,* leaf rust on wheat caused by *Puccinia recondita* f. sp. *tritici,* and root rot on soybean caused by *Fusarium oxysporium.*

The mechanism of action of YIB has been subject of considerable research. Studies indicate that YIB strains can colonize, migrate and multiply both inside and outside plant tissues. Twenty-four to 48 hours after germination of rape seeds inoculated with antibiotic-resistant mutant of YIB strain 83-10, up to 10^4-10^5 cfu could be recovered from individual roots. The bacteria, inoculated in form of strains 83-10 or A-47 on the seeds of cabbage, maize and wheat, were found to be able to migrate to and colonize also the stems and leaves reaching a population of 10^2-10^8 cuf per leaf. Yu et al. sprayed YIB strain A-47 on the leaf surface of wheat and the number of inoculant multiplied rapidly after spraying, reaching very high numbers around the stomata, wounds and sunken sites according to scan electronic microscope observations, antibiotic double marker test (rifampicin, streptomycin resistance), and leaf surface replica test.

The spore-forming *Bacillus* spp. have metabolically predominance for long periods and occupy ecological niches. They compete with pathogens for space and nutritional resources and have the ability to suppress, pathogen activity. The population of DRB strains Bacillus 112 and 114 greatly diminished on the rhizoplane of rapeseed after seed-soaking with YIB strain 83-10 suspension.

Liu indicated that application of YIB strain could eliminate the effect of some pseudomonads (DRB) in the rhizosphere of turnip. By spraying the suspension onto wheat leaves in the field and then isolating leaf samples weekly, the average numbers of *Cladosporium* spp. and *Alternaria spp.* on the surface of the leaves was reduced by 53.9% and 66.4%, respectively, compared to non-treated control samples. Some strains of YIB can produce antibiotics and inhibitory substances. A liquid culture of YIB strain A-47 inhibited spore germination in *Puccinia striiformis* f. sp. *tritici*, and *Cladosporium spp.* by 71.94% and 87.32% respectively.

The YIB culture also inhibited germ tube growth. Thus, YIB may reduce the incidence of infection by these fungi. Some of the YIB strains can secrete concentrated amounts of superoxide dismutase (SOD), an enzyme related to resistance, disease and aging stress. The SOD activity of rape (*Brassica napus*) tissue was found to increase after rape was treated with YIB filtrate. The activity of peroxidase, which is related to disease resistance, also increased after treating wheat plants with YIB. Enzymatic analysis showed that YIB strains also contain other enzymes, such as amylase, proteinase, and alcohol dehydrogenase.

The culture filtrate of YIB contains phytohormones, such as indole acetic acid (IAA), zeatin and gibberellin (G3). Peng and Xiong showed that YIB strain grown on^{14} C-glucose medium, could secrete IAA, G3 and zeatin. Anaerobic digestion is an effective method for treating many organic wastes, especially animal wastes and organic effluents from the food processing industry. Diverse anaerobic bacteria participate in the degradation process and produce intermediate metabolites as well as endproducts such as methane, carbon dioxide, and ammonia. The digested organic waste or slurry still contains organic matter, nutrients as well as microorganisms, and is superior over taw (undigested) slurry in terms of enhancing plant growth and improving the soil structure and nutrient status of soil.

Methane is one of the major products, of the anaerobic digestion process as a fuel in the form of biogas. About twenty years ago, Rural Biogas Programs were developed and implemented and biogas digesters were built almost everywhere in the Chinese countryside; most of them are small and intended for household use, few of them being larger for use by the whole village. By 1990, 4.7 million digesters had been built with a production of 1 billion cubic meter of biogas per annum.

The farmers make use of the biogas for cooking, beating and lighting. Biogas use can save labor for cutting firewood, protect trees and groves, prevent the burning of crop residues thus increasing organic matter in soil, and generally improve farmers' living condition. With time more advantages of anaerobic digestion were found.Digested slurry from biogas production is used comprehensively for fishing, cultivating mushrooms, and feeding livestock. It can increase crop output, augment farmers' income, decrease pollution, and hence attracts more farmers to use the technology. Anaerobic digested organic waste can be used to increase plant resistance to stress and diseases. For example, spraying supernatant of digested slurry on the leaf surfaces of orange trees can prevent freezing injury.

During the period of December 27-29 of 1991, temperatures in Jiangxi Province dropped to an usual low of— 7°C. Leaves of all Wenzhou tangerines in the area fell down because of the freezing damage and some of the trees died, except for one orchard run by a farmer who had sprinkled the surface of the leaves with the supernatant of digested slurry. In next spring, the fruit bearing rate in that orchard reached 94.7%, while only 45% in the untreated trees.

It has been shown that digested slurry can also increase cold-resistance and prevent decay of early rice, increasing the seedling survival rate by 8-15%. Compost has been successfully used to suppress certain soilborne diseases. Digested slurry can decrease the incidence of several plant diseases, such as vegetable (cucumber, tomato, watermelon) and cotton wilt diseases caused by *Fusarium oxysporum*, cotton anthracnose caused by *Glomerella gossypii*, rice blast caused by *Pircularia oryzae*, rice sheath blight caused by *Rhizoctonia solani*, black rot of sweet potato caused by *Ceratocystis fimbriata*, rice seedling rot, and barley yellow mosaic diseases.

Experiments in pot tests showed that a mixture of digested manure and soil decreases the incidence of watermelon fusarium wilt. High amounts of digested manure applied to soil had 86% control effect. Digested solid residues applied to early rice seedling beds reduced rice bud rot by 52.5% and seedling rot by 24% comparing to the application of chemical fertilizers (control treatment). After transplanting, rice seedlings turned green and tillered 2-3 days earlier than the seedlings treated with chemical fertilizers. Soaking seed with anaerobic digested slurry increases germination power and germination percentage, and promotes seedling growth.

The application of supernatant of digested slurry was found to be effective in controlling sucking pests, such as aphids, spider mites and planthoppers. The supernatant of digested slurry mixed with 450 g of Rogor emulsifier (40%) effectively controlled wheat aphids, as did 1500 a of Rogor emulsifier with water. Applying digested slurry can increase the quality of food crops, for instance by increasing the sugar content of fruits and vegetables, and improving appearance and taste.

Little is known about the mechanisms underlying the various effects of anaerobic digested slurry. Besides the metabolites mentioned above, digested slurry contains plant hormones such as indole acetic acid (1AA), cytokinin, and gibberellin. Zhu and Wei studied the mechanism of enhancing cold resistance of rice seedling and showed that digested slurry can decrease ion oxosmosis of cells, increase free proline content of seedling, moderate the decrease of ascorbic acid and peroxidase, and increase the relative content of unsaturated fatty acids, factors and processes known to be related to increasing cold resistance. Inhibition tests in petri dish showed that digested slurry strongly inhibits the growth of the plant pathogens *Heblminthosporium, sativum* and *Rhizoctonia solani,* rid moderately inhibits the growth of the pathogens *H.turcicum H. maydis, H sigmoideum* and *Colletotrichum gossypii.*

Shen found that ethylene and methane produced from digested slurry in microenvironments inhibited the activity of *Polymyza gracimimanis* responsible for the spread of yellow mosaic virus. Li et al suggested that the control of sweet potato soft rot in the storage period depends on the presence of the ammonia ion of digested slurry. Ma and Guo compared the physiological similarity of digested slurry and fulvic acid, a low aggregated humic acid.

4

Evolution of Bowerbirds

Bowerbirds occur across the Australio-Papuan region and are unique in that males build structures on the ground called bower, that appear to function inmate attraction and related activities. The first 7 years of my work on bowerbirds focused on the satin bowerbird (*Ptilonor-chynchus violaceus*) Previously, there had been no detailed quantitative study of any aspect of bowerbird behaviour. As an example demonstrating the value of comparative studies involving a large number of species, this work shows that even among closely related species there can be very large differences in behaviour.

One cannot characterise a group like bowerbirds based on studies of single species, and there is no typical species that fully represents this group. Information from numerous species often is needed to understand the evolution of complex traits like bower-building. Mate selection in species with elaborate male display traits was a topic central to Charles Darwin's seminal writings on sexual selection. Prominent in his discussion of sexual selection were the bowerbirds and their unique behaviour of building bowers. Bowers typically are made of sticks. In some species, bowers can reach 1.5 m high and are built near display court decorated with more than 2,000 decorations. These elaborate decorated structures frequently have been described as one of the wonders of the animal world.

There are now numerous hypotheses explaining how extreme displays evolve, although clear answers remain elusive. "Good genes" models propose that extreme sexual displays function as indicators of male quality to females choosing mates posits that female preferences can produce greatly elaborated male display traits without providing enhanced vigour to offspring. Other models include: passive choice, intrasexual signalling, proximate benefits and innate preferences. Recent empirical studies have shown that elaborate displays in polygynous species are typically not single traits, but a complex sets of traits.

In bowerbirds, bowers and decorations are part of a generalized display that includes plumage, acoustical and dancing elements directed at females during indicate that females prefer males with well-built and well-decorated bowers indicating that an important current function of bowers and decorations is to attract mates.

The large-number of models for explaining elaborated display traits, the lack of specific predictions from some of these models, and the complex array of traits involved in the display of polygynous species have made it difficult to develop clear-cut conclusions about the evolution of extreme male displays that bowers may function in female assessment of male quality as sires as a protective device for females being courted or as a structure with no direct functional significance with no direct context of sexual selection. Comparative studies of the evolution of traits among related species can provide critical information about the sequential evolution of the components of display and their initial and derivative functions. Several recent developments have made it possible to carry out a detailed comparative analysis of the function of bowers.

— The use of remote-controlled cameras aimed at bowers where males display and mate has allowed my students, field assistants, and me to simultaneously monitor all activity at as many as 30 bowers through an entire mating season. Because bowers are widely separated and the mating periods may last several months, it would have been impossible otherwise to obtain detailed information from numerous bowers.

— The advent of molecular techniques has made it possible to build a phylogeny of bowerbirds based on mitochondrial DNA sequence information.

The independent derivation of this phylogeny makes it possible to infer the historical pattern of the evolution of display traits by using maximum in parsimony to map the distribution of particular male display traits and their occurrence in ancestors onto the phylogeny. By combining these techniques, my students, collaborators, and I have been able to gather detailed information on the display traits of bowerbirds and map this information onto an 'independently derived phylogeny of the bowerbirds.

Building bowers is restricted to the family Ptilonorhynchidae is madeup of six genera. One genus, the catbirds, is monogamous and does not clear a display court. Members of the remaining five generations are polygynous, clear a court, and typically build a bower. There are two major designs of bowers.

Maypole bower-builders decorate a sapling with sticks. Males of species in the genus *Amblyornis* typically spire surrounded by a mossy circular court. and species build a hut-like dome that covers part of the court. *Prionodura newtoniana* build a two-spired structure with a cross perch connecting the spires. Avenue-builders build a two-walled structure with a display court near the end of the bower. Two other species, toothbilleds and Archbold's, clear display courts but do not build a characteristic bower. A phylogeny based on sequences of mitochondrial, DNA from the cytochrome b gene, developed in cooperation with Kusmierski and Crozier, indicates that there was an initial separation of lineages that led to the monogamous catbirds and the remaining polygynous species.

Later, a separation in the clade that led to bower-builders produced a lineage that led to the avenue-builders, and another that produced the maypole-builders and the two polygynous species that do not build bowers. Because building bowers is the dominant pattern in both lineages and it occurs in no other avain family, it most likely evolved once, preceding the split of the ancestors of the avenue and maypole-builders.

Hypothesis for the evolution of bowers

Many of the models of the evolution of extreme displays to develop specific models for the origin and evolution of bowers. For example, the placement of sticks in an incipient bower could have the result of an arbitrary female preference. Such a hypothesis, however, posits no particular functional role for bowers apart from mate attraction and fails, by itself, to account for the variation in form of bowers.

Evidence of a function for bowers, especially if it is consistent among different types of bowers, would weigh against this hypothesis. Another possibility is that bowers could have functioned to provide protection from various threats. Sources of threat include predators that might attack males and females, marauding males that force copulations on females while they view the court owner's courtship, and court owner himself forcing copulations on females not ready to mate. The hypotheses of predation and marauding-males are not well supported.

There is no evidence of predation on females or males while males are displaying on courts in more than 100,000 hours that cameras monitored the display courts of males. This is especially relevant because, in most species, males are not protected from predators by the bower during courtship. In addition, females are not protected in some types of bowers, including the open bowers of MacGregor's bowerbird, which may be similar to ancestral bowers, and those of the streaked bowerbird where the male occupies the covered part of the bower during courtship, opposite to what is predicted by the female protection hypothesis.

Last, females change their behaviour from being very reluctant to stand outside the bower before copulation to being quite willing to stand there afterward. The observations that many types of bowers do not confer protection, that the sexes stand in the wrong place, and that the females' behaviour changes after copulation also fail to support the hypothesis that provide protection from marauding males. The remaining hypothesis, protection from the courting male, is not limited by these difficulties.

Species from the two clades that build bowers show generally similar patterns of how they use bowers during Courtships that lead to copulations. In both lineages, males display facing the female. The bower enhances the female's ability to escape unwanted mating by blocking the male's direct path to her and forcing him to run around the wail of the bower or maypole to mate.

The increased distance that males must travel to reach a female enhances her ability to escape unwanted matings. A plausible solution is that building bowers may offer a compensating benefit that outweighs losses from forced copulation. If the reduced threat of forced copulation causes more females to be attracted on courts with bowers than those without, then gains from increased visitation by females could more than compensate for the lost ability to force copulations with females. Second, how do we test the hypothesis that females prefer males with bowers that function to protect them from forced copulation. It is rather tricky to infer the function of a trait at its evolutionary origin, given no fossils or other record of the ancestral form of the bower. Information from observation of the different species of bowerbirds and knowledge of their evolutionary relationships can be used to establish likely hypotheses.

The criteria that would strongly support a hypothesis for an incipient function of bowers are:

1) The proposed function should be consistent with the design of the likely ancestral bower

2) This function should be significant across all types of bowers, and

3) Species that do not build bowers should show alternative solutions to the problem solved by the bower. There is no guarantee that even the correct answer will meet all three criteria. It is possible, for example, that bowers could have taken on a variety of secondary functions that have replaced the incipient function of the bower.

Display in the satin bowerbird

Satin bowerbirds are representative of species that build avenues. They occur in rain forests that fringe eastern Australia. The bowers of satin bowerbirds have two walls of sticks separated by a central avenue where females stand and then crouch when they are courted.

The bower is aligned in a north-south direction with a decorated display court on the north end. The male displays while facing the female from the display court with a decoration held in his beak. The females visiting the bower and the court typically alight in the vegetation on the south side of the bower and then move directly into the avenue of the bower. Initial vocalisations consists of numerous guttural chortles and squeaks that progress into a typical call sequence that consists of an initial mechanical buzzing followed by mimicry of a kookaburra, a Lewin's honeyeater, and less frequently, a crow. The buzzes in the mechanical calls occur in conjunction with rapid movements by the male across the north entrance of the bower accompanied by flicks of one or both of his wings. Females signal their willingness to copulate by lowering from an upright stance to a crouch. To copulate, the male circles to the opposite side of the bower to mount the female.

After a brief copulation, the female shakes intermittently in or near the bower for up to several minutes and then leaves. Females usually visit the bowers of several males, but mate with only one. The average courtship lasts slightly more than 4 minuts. In satin bowerbirds and other avenue-builders most courtships end with the female leaving from the north entrance of the bower as male moves from the court toward the south entrance in his attempt to copulate.

Only 90% of satin bowerbird's courtships are successful, although the most attractive males mate in 25% of their courtships. There is a significant relationship between the number of decorations and the mating success of males, indicating that it is important for females to enter the bower and see the display in order to assess the males.

Display in macgregor's bowerbird

Macgregor's bowerbird occurs at high elevations in the mountain ranges across central and eastern New Guinea is bower is a simple maypole, a sapling decorated with sticks and moss. Commonly, the sapling is rather thin and sticks are placed nearly horizontally increasing the diameter around the sapling to approximately 25 cm. The low part of the maypole and the floor of the round court surrounding it are covered with a fine compressed moss mat that rises to form a circular rim approximately 40cm from the maypole. Many small decorations are used on the court, including insect parts and seed pods.

Regurgitated fruit pulp is hung near the ends of the sticks of the maypole bower. Numerous large woody black fungi are arrayed on the rim of the court and on nearby logs. Females arriving for courtship often land on the maypole, move down it, and then hop onto the court. The male usually is already present on the court and may have been calling prior to the female's arrival.

The male positions himself on the opposite side of the maypole from the female with his chest pressed up against it and with his head plume concealed. He calls and, as the female moves around the maypole, he makes a counter move so as to keep the bower between him and her. Calling and counter moves continue for approximately 4 minutes before the male increases the intensity of display by expanding his bright orange head plume and violently shaking his head from side to side. The side-to-side shaking is associated with a rapid foot movement that appears to counterbalance the rapid movement of his head.

See from the female's side of the bower, this display creates rapid orange flashes on each side of the maypole. After several bouts of head-shaking the male moves around the bower toward the female in an attempt to copulate with her. Although the shape of avenue and maypole bowers is fundamentally different, there is a striking similarity between maypole bowers and those of avenue-builders in the way in which the bower is used to separate the male and female, in each case, the male develops a prolonged courtship display.

Display in toothbilled bowerbirds

Toothbilled bowerbird courtship is very different from that of other species of bowerbirds. Males clear courts but do not build bowers. Male courts are close together and aggregated into a lek. Males interact with frequent loud calls, with dominant males interrupting the calls of males on adjacent courts. Unlike other bowerbirds, they use large objects as decorations on their Courts.

Courts surround several trees of small diameter, the bases of which have been cleaned meticulously of debris. Females arrive on the court and stand very still, as if waiting for the male. Toothbill courtship is very brief. The male aggressively mounts the female with little or no display after she arrives on the court. The average time of courtship is 3.8 seconds.

However, copulations are prolonged and violent relative to the brief and more cooperative behaviour seen in other species. During these copulatory bouts, the male continues to display with characteristic low buzzing calls and wing beats. Females leave the court immediately after mating without the prolonged flapping of wings that is characteristic of other species. By far, toothbill males spend less time on their court than any other species, and it appears that the adjustments in their display and mating behaviour reflect an especially high susceptibility to predation while, they arrive on the ground.

The aggressive nature of the courtship suggests that females might not be fully prepared to copulate when they arrive, but the need to reduce time on the ground has caused males to attempt to speed this process. The loud vocal interactions of males on courts and the use of large decorations suggest that females can evaluate male qualities such as dominance before they arrive on the court. At lek centers, males that preliminary studies show are dominant in vocal interactions have the highest mating success.

The correlation suggests that females are choosing mates. Observations by Frith and Frith and our group show that males hiding behind a tree on their court sometimes call to females on nearby perches. However, we have not seen this calling lead to copulation. In toothbills, if male calling interactions and large leaves allow females to choose mates before they arrive on the courts, then bowers may not be necessary.

The female already may have selected the male before arriving at his court, so the prospect of forced copulation is not threatening. The capture of females by males indicates that forced copulations possible in bowerbirds. The hiding of males behind trees during calling displays suggests a situation analogous to the initial condition of bower evolution in which females seek some protection from the courting male.

Archbold's bowerbird

Archbold's bowerbird (*Archboldia papuensis*) is the other polygynous species that does not build a bower. Male Archbold's clear a large display court overlain with a mat of ferns where males place a variety of decorations, including snail shells, dark fruit, beetle wings, arid King of Saxon Bird-of-Paradise (*Pteridophora alberti*) head plumes. Typically, small decorations are in piles near the fringes of the display court and are arrayed on the limbs overhanging the court.

Male Archbold's bowerbirds neatly drape limbs that cross up in above the display court with a uniform curtain of flowerless orchid vines that nearly touch the display court and subdivide it. The cumulative array of the curtains provides a rather dramatic visual effect. Male Archbold's bowerbirds are large and uniformly black, with a bright yellow head crest that extends from between the eyes, over the top of the head, to the neck.

The crest covers more area on the bird's head than the crests in species of Amblyornis, but the individual plumes are much shorter. Courtship in Archbold's bowerbird also is unique. Males chase females around the court. Occasionally the female pauses, and the male stops near her and attempts to approach her with his body pressed close to the court. In Archbold's bowerbird, courtship is not constrained particular site, as it is hi species with bowers.

The preliminary phase of courtship about the court, and the large size of the of the court may accommodate these chases. The walls may function in constraining the direction females can move. Male Archbold's bowerbirds have evolved an alternative solution to the problem of copulating with females in the absence of a bower, Unlike toothbills, they do not interact over long distances or have leks, and they do not attempt to grab and copulate females by force. Like most other bowerbirds, they have prolonged that choice occurs on the display.

The low position males during courtship and frequent female rejection of the display court. The low position of males during courtship does not compromise the female's ability to escape the courting male, even when lie is nearby. One explanation for the loss of bower-building in Archbold's bowerbird may be related to the widely ranging displays that females use to test males, In most species of bowerbirds, males can be directly compared because they compete by stealing their competitor's decorations and destroying their bowers.

Visiting females may assess male competitive ability by the quality of his display. In Archboldia, the bowers are spread very far apart, and the possibility of male interaction is low. The frequent chases may be a means by which females test male athleticism. In *Archboldia*, where the bowers of males are spread apart in small forest islands, there may be a similar need for males to demonstrate fitness in athletic, rather than interactive components of display.

Evolution of bower

The hypothesis that protection from the courting male is important in the evolution of bowers is supported by the patterns of evolution in bowers and its consistency with the diversity of types of bowers. The ancestor to the lineage that led to modern species that build bowers probably displayed off a decorated ground court. Females favoured courts with a natural barrier, such as sapling, that separated them from males during courtship because it allowed them to approach the male and closely observe his display and decorations while still retaining the ability to leave if not stimulated by the display. Males who enhanced this barrier, *e.g.*, placed sticks around the sapling and enlarged its diameter, offered females a safer vantage point for observing display. Males could gain from this elaboration by exploiting the female preference for mating in a protected environment.

Increased female visitation, and lessened threat during courtship contributed to an overall increase in matings over what might be achieved by forced copulations. Gains for females from the avoidance of forced copulations might include eschewal of genetically inferior males and reduction of direct physical cost. Remating of females which have been forced to copulate other males would lower the value of forced copulations to court owners have caused an males to shift efforts toward attraction of females. The simple bower described above is similar to the bower built by MacGregor's bowerbird. Once the tendency to use sticks to build a bower evolved, however, it was possible to build a bower that, functioned in female protection but did not require a central sapling.

The loss of dependence on the use of saplings could have allowed males more freedom in selecting sites for bowers, in orienting their decorations, and in displaying their decorations. The ancestors of avenue-builders probably added a second barrier because it oriented females toward illuminated parts of the court where males could concentrate their decorations on a well-lit stage.

The orientation of bowers in a north-south direction, the consistent placement of most decorations on the north side, and the clearing of leaves over display courts support this hypothesis. This hypothesis is consistent with all types of bower-construction and our observations of how bowers are used in courtship. In both avenue and maypole bowers, males are forced to run around a barrier in order to reach the rear of the female where they can copulate. The delay caused by this extra travelling time gives females an opportunity to escape males that are unattractive.

The behaviour of the two species that do not build bowers also is consistent with the protection hypothesis. In toothbilleds, there is no bower but because females appear to move to the ground for copulation only after they make their mating decision they leave the court without mating proportionately less often than females in other species.

5

Biodiversity in Neotropical Primates

Traditional classifications divided New World primates into "marmosets" or "callitrichids" and "nonmarmosets" or "cebids," in which the marmoset patterns of morphology were seen as primitive and those of cebids as advanced. Most authorities recognize 16 extant platyrrhine genera, including Ateles, Brachyteles, Lagothrix, Alouatta, Pithecia, Chiropotes, Cacajao, Aotus, Callicebus, Cebus, Saimiri, Cebuella, Callithrix, Leontopithecus, Saguinus, arid Callimico.

In contrast with the general concordance of reviewers working at the genus level, considerable disagreement exists among authorities working at higher taxonomic levels. In a 1986 revision, Ford reorganized recent platyrrhine genera using a cladistic approach. In 1992, Rosenberger of the National Zoological Park, Smithsonian Institution, launched a new synthesis based on phylogenetic relationships, adaptive change, and the fossil record. He recognized four monophyletic subfamilies in that classification, Callitrichids became callitrichines, closely related to squirrel monkeys and capuchins. Many features of callitrichines are now thought to have resulted from selection for small size and thus are derived and not ancestral or primitive traits. Identification of the phylogenetic link between the Cebinae and Callitrichinae was made possible by the discovery of a new set of fossils at La Venta, Colombia, with morphological characteristics intermediate between those of the two subfamilies.

The rate of description of fossil platyrrhine genera has increased markedly in recent years. Of 16 known fossil genera, 7 were described since 1985 and 4 of these since 1990. Our understanding of phylogenetic relationships and extinction of neotropical primates certainly will continue to improve if the trend toward increased description of fossil forms continues. New advances in molecular genetic techniques now make it possible to objectively test hypotheses about the phylogenetic relationships among the platyrrhines.

The Brazilian geneticists Maria Sampaio and Horacio Schneider examined sequences of DNA in nuclear globin genes and found close relationships between Alouatta and the atelines, between Cebus and Saguinus, and among Pithecia, Chiropotes, and Cacajao, adding additional support to the phylogenetic model developed by Rosenberger. Much of what we know about the diversity of neotropical primates, at the species level is due to the painstaking descriptive detail provided in the published works of Hershkovitz.

That author and a few others generally used pelage and cranial descriptions to classify the majority of extant species of neotropical primates. However, there been considerable recent controversy about the criteria used in assigning species status to many of these forms. Geographic ranges of many forms have been altered by recent anthropogenic habitat modification. As a result, correlations between patterns of geographic distributions and morphological characteristics become increasingly difficult to interpret.

The biological accuracy of taxonomic designation becomes particularly important because legislative protection and conservation funding for full species is significantly greater than those for subspecies. In addition, conservation objectives of reintroduction and translocation of isolated populations are predicated on the conservation of genetic diversity within a species. If genetic management is to have imperative that we be able to distinguish conspecifics forms less closely related forms.

Revision of neotropical primates at the species level should be given high priority, both in terms of contributions to primate conservation and basic search. Recently developed noninvasive genetic techniques such as sequencing of mitochondrial DNA extracted from hair follicles of museum may make it possible to resolve some of these controversies. Living neotropical primates are found from central Mexico to northern Argentina and include some 68 species, comprising about one-third of all living species of primates. In the mid 1980s, much emphasis was put on identification of major phytogeographic regions that contained relatively large numbers of primates, particularly rare, endangered, or endemic species. These analyses resulted in maps of priority areas for primate conservation and research. The discovery of four unknown primates in such a short period of time, including one species in the backward of one of the largest cities in the world, underscores how little we really know about the diversity of even well-studied taxa such as nonhuman primates and emphasizes the need for more basic research and surveys on primates in the Neotropics. However, if conservation efforts on behalf of any species of primates are to succeed, it will be necessary to go beyond the identification of its geographic distribution and taxonomic relationships with other primates. Although biologists have been conducting censuses of neotropical primates for decades, lack of standardization of methods and the difficulty of observing primates in closed forest has made it difficult to assess the accuracy of the results of these surveys.

Nonetheless, numbers of neotropical primates appear to be decreasing. For example, in 1969, the Brazilian conservationist Alvaro Aguirre estimated that 2,000-3,000 muriquis survived in Brazil's Atlantic forest. In the mid-1080s, the minimum number of muriquis known to exist in 10 locations as published as 240. However, it was not until 3 years ago—when Kicrulff, a Brazilian graduate student, used playbacks of golden lion tamarin vocalizations in virtually every forest that might contain these primates-that an accurate estimate of the number of individuals in the wild became known.

This, the first exhaustive census for any neotropical primate, suggested that only 559 individuals in 103 groups remain in the wild and that the species is subdivided into 16 isolated subpopulations, 12 of these containing only 1 group. The accurate nature of this information removed any question about the degree of threat to this species arid prompted significant changes in the focus of conservation and related research activities.

Monitoring changes in quantity and quality of habitat

In 1986, critical regions for conservation of neotropical primates were identified as huge phytogeographic areas such as the Amazon basin and "megadiversity countries" that contained a large diversity of primate taxa. Since that date, our ability to pinpoint regions for conservation has improved markedly. In 1990, approximately 100 specialists on Amazonian fauna, flora, soils, and climate met in Manaus, Brazil, to produce a map of biologically important regions in the Amazon Basin based on criteria of endemism rarity, and diversity.

Important here is the diversity of specialists contributing information and the speed with which the results were disseminated. Perhaps the most significant recent advance in our ability to quantitatively evaluate primate habitat is the combination of remote-sensing technology and geographic information system (GIS) analysis. A decade ago satellite images of regions of South America were expensive, difficult to acquire, and could be analysed in only a very few locations.

Currently, a variety of high-quality inexpensive magnetic image data are available from several sources and cover virtually every region on Earth. Recent improvements and reduction of hand-held geographic position system (GPS) equipment now allow a researcher to precisely determine latitude and longitude for samples of habitat, thus simplifying the ground-truthing of satellite data. Desktop computer versions of GIS software make it possible to conduct sophisticated habitat analyses without acquiring specialized equipment.

New techniques such as gap analysis, which evaluate the protection status of plant and animal communities by GIS overlay of distributional data on maps of existing protected areas, help conservationists to be proactive rather than reactive in their efforts to preserve biodiversity. The accessibility of satellite data and new analytical techniques resulted in three large steps forward in terms of establishing conservation and research priorities in the Neotropics.

First, we now are able to quantify changes in vegetation patterns over time. For example, in tire 1980s, widespread disagreement over rates of deforestation in the Brazilian Amazon hampered efforts at international collaboration in conservation. In the 1990s, independent reports by Brazilian and American scientists agreed that previous estimates were several-fold larger, than actual rates of deforestation.

Since future treaties on biodiversity and climatic change are likely to require countries to limit their emissions of greenhouse gases, including those resulting from deforestation, estimates of deforestation based on satellite data acquire additional political and economic importance. Second, we now can use satellite data to test hypotheses about the effects of anthropogenic factors on habitat quantity and quality. For example, we can use archival data to document habitat changes that followed the construction of a road into a forested area in the Amazon and subsequently use that information to predict how a proposed road might alter undisturbed forest in another region.

Third, based on any criteria that can be mapped, it is now possible to quantitatively evaluate areas for creation as reserves, or as corridors between reserves. Remote sensing and GIS techniques are becoming increasingly powerful research tools and should be included in the tool kit of any researcher or conservationist doing field work on neotropical primates.

Ecology and evolution

Hundreds of scientific reports have been published in the past decade on the ecology and behaviour of neotropical primates. Field research using the comparative method, long-term demographic studies, and research addressing the effects of human disturbance factors on primate population size and structure. Although the principles of evolutionary biology should be useful ill explaining primate behaviour, primates rarely have been at the forefront in this work. Cheney *et al.* identify several reasons why primates are rarely used to test evolutionary theory.

Most species of primates are difficult to identify individually or even to observe under field conditions. Many species are so long-lived that accumulation of data on life histories may require decades of observation. In addition, the social behaviour of many species is sufficiently complex that it becomes difficult to collect information on a sufficient number of individuals to obtain a complete picture of patterns of social organization.

Seasonal and annual variation further complicates the design of field studies examining the effects of ecological variables or the adaptive significance of specific behaviours in tropical ecosystems. Finally, formulation of generalizations often is impeded by differences in researchers, methodology. One approach that has proved useful in circumventing these obstacles is to ask how several primate taxa behave under the same circumstances and when the same observational methodology is used.

Conclusion then may be inferred from differences between species. This comparative approach has the added advantage of focusing on an interesting question rather than on an interesting primate. A good example of the comparative method applied m field research on primates is Terborgh's long-term study of the comparative ecology of five New World primates in Manu National Park, Peru.

Comparisons across species allowed Terborgh to make inferences about, for example, the function of territorial defence, factors that influence size of the group and the relationship between size of the group and the social system in primates. Another fruitful comparative study was conducted by Peres at a remote site in Amazonas, Brazil. Peres compared the feeding ecology and habitat use of 13 primate species for a period of 20 months.

The comparative method allowed Peres to conclude that seasonal migrations by several species living in large groups were the result of "bottlenecks" in food availability in their habitats. An interesting example of the comparative method is the comparison of wild- and captive-born reintroduced golden lion tamarins on private property adjacent to Poco das Antas Reserve, Rio de Janeiro, Brazil.

Parallel studies of native and reintroduced tamarins allowed researchers to generalize about the effects of captivity on locomotion, communication range use foraging, and reproduction. A second approach that has proved useful both in terms of basic research and conservation application is long-term monitoring of derflographic variables for a single primate population.

During the past decade of study on golden lion tamarins in Pogo das Antas Reserve, researchers individuality marked 80% of the tamarins in the reserve and these became habituated to the presence of human observers. Continuous monitoring of all births, deaths, emigrations, and immigrations in 22 study groups allowed the researchers to compare average fecundity and survivorship for tamarins following specific behaviour patterns.

The conservation impli-cations of these data are discussed below. Calculation of reproductive success under field conditions can be problematic, particularly for species of primates in which more than one male copulates with a reproductive female. Under these circumstances, new molecular genetic techniques, notably DNA finger-printing, may allow paternity exclusion and thus direct determination of male reproductive success. It long has been thought that deforestation and have negative effects on most primate populations. However, the effects of these and other anthropogenic factors rarely have been quantified for neotropical species.

In recent reviews of the impact of habitat disturbance on a variety of species of primates, Johns and Johns and Skorupa reported" that body size and degree of frugivory were negatively correlated i species' probability of survival. Redford and Robinson surveyed the available literature and predicted that hunting by humans would cause primate population density to decrease as prey body mass increases. Freese *et al.* surveyed primate densities in Bolivia and Peru and suggested that predation by humans was the most important factor affecting primate densities.

Peres used transact censuses of primates in hunted and unhunted areas to test hypotheses about the effect of hunting cn primate densities. Peres found that body size alone largely determined the choice of primates hunted in Amazonian forests and that large species, such as woolly monkeys, often were hunted to extinction in areas near human settlements, even in the absence of habitat disturbance. Redford reached similar conclusions for a variety of large animals in neotropical forest. Two conclusions are suggested by these studies.

— Conservation efforts need to focus on large-bodied primates with low rates of reproduction, even before habitat destruction begins.

— Past and present hunting pressure must be included in population dynamics of large-bodied primates, even in areas of apparently intact habitat.

Conservation strategies

The alarm that sounded in the mid-1980s about the immediate threat to most species of neotropical primates was answered in a number of ways. I have classified these strategies as information management and networking application of scientific method, captive breeding, conservation education and inter national collaboration. The advent of widespread use of electronic mail dramatically increased the speed and decreased the cost of international communication.

Transfer of data files, even from remote field stations, has become routine in the past years. Computerized bulletin boards are available on a number of relevant topics, including conservation biology, primates, ecology, the Neotropics in general, as well as several tropical countries. Many international nongovernment conservation organisations (NGOs) support regional programs to develop and conduct conservation and activities in neotropical countries.

Some of these initiatives, e.g., the Golden Lion Tamarin Conservation Program in Brazil , which has been supported for 10 years by grants from the World Wildlife hind and other NGOs, focus on conservation of primates and their habitat. The Nature Conservancy sponsors the establishment of Conservation Data Centers in 13 Latin American and Caribbean counties to aid in decision-making in conservation and sustainable development. Conservation International supports a Rapid Assessment Program in which teams of experts census diverse taxa of fauna and flora to set priorities for conservation action.

Scientific method

The past decade has seen a heartening increase in the generation of conservation theory and the application of rigorous scientific method and problem-solving techniques to conservation questions. Graduate degrees in conservation biology now are offered at many universities in the United Stales and Europe.

The National Science Foundation now funds grant proposals in conservation and restoration biology. Although the primary concern of conservationists is the survival of ecosystems, much of the theoretical and applied research in conservation biology is conducted at the population level. The link between population and ecosystem conservation is thought to be the appropriate selection of population of "flag ship species" that serve as the focus of conservation campaigns, "umbrella species" that have ranges large enough to include significant portions of natural ecosystems, and "keystone species" whose presence or absence determines the abundance and distribution of a number of other species.

Appropriate selection of populations for conservation can conserve biodiversity effectively. The charismatic nature and ecological role of many neotropical primates makes them well-suited to be flagship species. A new approach to determining conservation action plans at the population level is the Population Viability Analysis or Population and Habitat Viability Analysis. This concept is predicated on the need to establish quantitative conservation goals for the population under study, e.g., a 95% probability of survival over a period of 100 years.

Computer models such as VORTEX are used to estimate the probability of population extinction and time to extinction and time to extinction, based on the effects of stochastic events such as inbreeding, environmental variation and catastrophes on population demographic variables. The probability of survival of multiple subpopulations in a metapopulation also may be estimated.

Examples of PHVA applied to four species of lion tamarins are presented in Seal *et al.* The advantages of PHVA include the simultaneous treatment of several variables, the collaborative process by which the best available information is assembled for all interested parties, the quantitative nature of the results, and the versatility in evaluating management options. The shortcomings of PHVA are related to the assumptions implicit in the computer model and the effects of estimating unknown values.

For example, the impact of environmental variation on fecundity and survivorship rarely is known for population is nature. Small changes in the specified effects of environmental variation on reproduction or mortality can produce 10-fold differences the predicted probability of extinction of the population.

Captive breeding and reintroduction

Not long ago, zoos were seen as consumers of primates and other animals, competing to assemble a collection of the rarest and most valuable examples for public exhibit. Recent changes in the objectives of modern zoological parks make these institutions important contributors to in situ conservation of neotropical primates. The principal objective of captive propagation now is seen as the re-establishment or reinforcement, not replacement of wild populations. With over 13,000 specimens in zoological collections, primates are overrepresented relative to other mammalian orders.

Nevertheless, only about 50% of primate taxa are represented in captivity. International management committees, organized to maximize retention of genetic diversity and minimize probability of Ion in captivity and the wild, are increasing for many vertebrate taxa, including primates. A number of advisory groups work to coordinate the ex situ and in situ primate conservation activities of zoos.

The Species Survival Commission (SSC), consisting of world experts on approximately 100 taxa or conservation topics, is the main source of advice to The World Conservation Union (IUCN) regarding the technical aspects of conservation of species. The SSC promotes action on the part of the international conservation community on behalf of species threatened with extinction and those important for human welfare. Two SSC specialist Groups are particularly important for primate conservation: the Primate Specialist Group, comprised of about 200 experts working in the conservation of primates, and the Captive Breeding Specialist Group, whose objective is the conservation or establishment of viable populations of threatened species.

In 1992, a Taxon Advisory Group (TAG) to the American association of Zoos and Aquaria (AAZPA) was formed to coordinate the facilitate captive breeding of New World Primates in support of in sita conservation efforts. Faunal Interest Groups (FIGs) coordinate conservation activities in zoos for fauna and flora from a specific region, e.g., Brazil. Studbooks, containing complete demographic records for a species ill captivity, facilitate the scientific management of captive populations, e.g., by facilitating selection of appropriate individuals for pairings reduce loss of genetic diversity. Studbooks are maintained for 15 species of neotropical primates, and have been proposed for several others. For a few species, information on native populations is sufficient for inclusion in studbooks.

The reintroduction of captive-born neotropical primates has been initiated for several species in the past few years. In addition to the reintroduction of golden lion tamarins mentioned above, reintroduction and monitoring of *Callithrix geoffroyi* in Espirito Santo state, Brazil, began in 1991. The costs and criteria for success in a primate reintroduction are reviewed by Kleiman et al. The IUCN-SSC Reintroduction Specialist Group and the AAZPA Reintroduction Advisory Group both were created in the past few years to facilitate in situ conservation through integration of wild and captive population management.

Conservation Education

The role of community education in directing the attitudes and behaviours of local peoples toward sustainable use of natural resources long has been recognized as an essential component of any viable conservation strategy. This is particularly true where charismatic species such as primates can be used as flagships for conservation efforts. However, only in the past few years have rigorous evaluation techniques been used to quantify the effects of community education in primate conservation projects. Illustrating the value of quantitative evaluation of community education, educators conducted surveys before and after 2 years of activities related to the golden lion tamarin conservation project and documented a significant increase in the percentage of respondents who recognized golden lion tamarins in a photograph.

Whereas 42.5% of these responses were attributed to project activities, only 6.9% reported seeing tamarins in the forest were mentioned most often included television and radio, a total effect roughly equivalent to all other project activities combined. Graduate training in conservation education for South and Central American nationals is one of the best investments that can be made to ensure the future of neotropical primates.

International collaboration

A final example illustrates progress in international collaborations. In this case, private lands were purchased and annexed to a Brazilian federal reserve too small to guarantee the survival of an endangered primate. A PHVA for golden headed lion tamarins, conducted in the Una Biological Reserve, southern Bahia State, Brazil, suggested that the probability of extinction of that primate was relatively high. Based on the findings of the PHVA, a coalition of eight international conservation organizations conducted fund-raising efforts to buy out squatters and purchase land to annex to Una Reserve.

Funds were transferred in several instalments to Fundacao Biodiversitas, a Brazilian NGO, which purchased lands at market price from local landowners and immediately conferred title to the Brazilian agency responsible for administration of the Reserve. To date, the area of the Reserve has been increased from 5,342 hectares to 7,059 hectares.

6

Human-caused Extinction of Birds

Birds have played a prominent role in our understanding, of the processes by which species become rare, endangered, and finally extinct. The resulting literature on avian conservation biology has proliferated for decades and now is part of the information explosion, with all of its benefits and frustrations. Each year we read of additional species of birds whose existence no longer can be demonstrated. One of the latest is the Colombian grebe, *Podiceps andinus*, whose demise in highland Colombia is attributed to the loss of wetlands, the introduction of exotic fish, and hunting.

Declaring a species extinct can be a tricky business; the discovery of even one living individual, regardless of the long-term viability of the species, refutes the claim. In other words, negative evidence can be refuted by even one bit of positive evidence. To discover a few living Colombian grebes, however, is unlikely to prevent extinction of the species, given the vulnerability of very small populations to demographic stochasticity, catastrophes, genetic viability, and disease. Most species declared extinct by ornithologists never have been redisovered. One exception is the Cebu flowerpecker, *Dicaeum quadricolor*, endemic to the Philippine island of Cebu, where 8 of the 12 endemic subspecies of birds already are gone.

Considered extinct since 1906, up to four individuals of the Cebu flowerpecker were observed in 1992 and 1993 in the last remaining patch (<2 km^2) of closed canopy forest on the entire island. The prospect of long-term survival for the Cebu flowerpecker is remote, given the scarcity of its habitat and its small population. Only 2 years after its rediscovery, the Cebu flowerpecker may be gone for good. Extinction is occurring today at unprecedented rates across a brand range of terrestrial and aquatic habitats.

Because species richness of birds is so high in tropical forests, the single most prolific cause of endangerment and extinction in birds is the destruction of tropical forests. Deforestation and other types of habitat loss also deplete avian communities in temperate and high latitude areas, which typically have fewer species of birds to loss than tropical areas. We have no evidence that any of the species of birds now endangered or that have gone extinct in recent millennia would be in their predicament if not for human activity.

While extinction does occur naturally, human impact has increased rates of extinction by orders of magnitudes over background rates, and therefore is the only significant cause of our current "biodiversity crisis." A comprehensive attempt to set global conservation priorities for birds has been compiled by Bibby et al. who identify 2,609 species of birds with breeding ranges of less than 70,000 km^2, designating these as "restricted range species" (RRS). Sets of these species tend to occur together on islands or in well defined areas of a particular continental habitat, especially tropical or montane forests. An endemic bird area" (EBA) is any place where two or more RRS are sympatric. The 221 EBAs each contain 2-67 RRS.

Many of the EBAs correspond roughly or rather exactly with centers of endemism in other organisms such as vascular plants, butterflies, amphibians, reptiles, or mammals. The total numbers of both EBAs and RRS are about evenly divided between islands and continents. The tropics have 76% of all EBAs and more than 90% of all RRS. The data on EBAs and RRS provide a rough but informative idea of the potential for extinction of birds in upcoming decades.

For example, the country with the most RRS is Indonesia, where most islands are unprotected. The EBA with the most RRS is the Solomon Islands were none of the land is officially protected. Of the 10 EBAs with the most RRS, only one has more than 15%, of its land under protection. We are rapidly approaching the point of no return for hundreds of species. Humans cause the extinction of birds in four major ways:

— direct predation;

— the introduction of nonnative species;

— the spread of disease; and

— habitat degradation or loss.

Direct predation includes hunting, gathering eggs, or removing nestlings for captive rearing. The introduction of nonnative animals exposes indigenous birds to new predators, competitors, parasites, or pathogens, and thus is related to categories (1) and (3), Habitat degradation or loss can be direct (deforestation, draining wetlands, plowing prairies, toxic pollution, etc.) or due to encroachment by nonnative plants.

Much of our biodiversity crisis is due to human impact of recent centuries, especially the past few decades. Cooperative research by archaeologists, geologists, and biologists has shown, however, that most plant and animal communities were not pristine in preindustrial times. Our skills with tools, and fire have set us apart from other animals for tens of millennia.

All human societies, even nonagricultural hunter-gatherers, have had various effects on their environment. Hunting, fishing, and gathering often focus on certain species or groups of species. Agriculture affects natural ecosystems in more diverse ways, including modifications of landscape, soils, and water supply through deforestation, erosion, channelling, flooding, draining, etc., as well as the elimination or propagation of selected species of plants and animals.

Table 1 Ten countries with the most restricted-range species(RRS) of birds.

Country	*Occurring RRS*	*Confined RRS*	*Threatened RRS*	*Number of EBAsa*
Indonesia	411	339	18	12
Peru	216	106	51	18
Brazil	201	122	67	19
Colombia	189	61	51	14
NewGuinea	172	82	18	12
Ecuador	159	32	38	11
Venezuela	120	40	17	8
Philippines	111	106	36	9
Mexico	102	,59	23	14
Solomon Islands	96	43	16	4

An understanding of past plant and animal communities is important for long-term estimates of community stability and change, such as how thoroughly and rapidly ecosystems might recover from disturbances. The responses of plants and animals to future climatic changes will be compromised by human-caused habitat fragmentation. Some species may be unable to disperse across tracts of disturbed habitat. When feasible, past distributions can aid in planning translocation programs, with an overall goal of preserving species assemblages that at least approach those of a less disturbed state.

Continents

The distinction between continents and islands is useful, although there is a continuum in land area and isolation from small, remote oceanic islands, such as Ascension or St. Helena, to large "continental" islands, such as madagascar, Borneo, or New Guinea, to a relatively small and isolated continent such as Australia. Regardless, we now may be at a threshold where human-caused extinction of birds—heavily biased thus far toward species on islands—will occur at comparable rates on continents.

Another important factor in endangerment and extinction, much more species-specific than habitat loss, is direct exploitation. This may take the form of hunting, or taking young as pets. The combination for habitat loss and direct exploitation has had verifiable impacts on one or more species from virtually every family of birds. Unlike on some islands, we know little about the prehistoric impact of people on birds on most continents.

A partial exception is North America, where people first arrived about 11,000 years ago. Between 20 and 40 species of birds became extinct at this time, probably because of ecological dependencies on the more than 40 species of ground sloths, mammoths, mastodons, horses, tapirs, camels, and other large mammals that died out across the continent. If humans were involved in the mammal extinction, then they indirectly caused the extinction of the birds as well. Before the large mammal communities collapsed, the carrion-feeding California condor lived as faraway from California is Texas, Florida, and New York.

Most other scavenging birds perished completely. In the next 10,000 years before the arrival of Europeans, however, only two North American birds)re known to have become extinct: a flightless duck, *Chendytes lawi*, of the Pacific coast and a small turkey, *Meleagris crassipes*, from the Southwest. Bones from late prehistoric archaeological sites in North America document birds such is the trumpeter swan, Mississippi kite, Swallow-tailed kite, whooping crane, sandhill crane, long-billed curlew, great auk, Carolina parakeet ivory-billed woodpecker, common raven, and fish crow in localities well outside of their modern ranges. While intertribal exchange of birds might account tor sonic of these range extensions, most seem to reflect former indigenous populations.

Prehistoric hunting, trapping, and habitat modification, as well as natural climatic change, may have caused these range contractions. It probably is no coincidence that three of these same species are now extinct and the whooping crane is endangered. Some birds, such as the California condor, bald eagle, and golden were hunted for feathers, bones, or ceremonial purposes more than for food. In the past 200 years, at least five species of North American birds have been lost. Each of these, except the woodpecker, represented a monotypic genus.

Two other species, the Eskimo curlew and Bachman's warbler, are either extinct or virtually so. The California condor, whooping crane., cockaded woodpecker, black-headed vireo, warbler, and Kirtland's warbler persist today only in dangerously small, localized populations. Without management, three of these species already would be extinct. Even if no more habitat is lost, some of these last eight species are likely to become extinct in the next 200 years, as may others that are known to be endangered, in decline, or localized today.

Losing 20-25 species per millennium may seem slow to politicians and economists, but is a devastatingly high rate of extinction from all evolutionary standpoint. We have no evidence that speciation will offset any of those losses. Long-term ecological data are lacking for most tropical localities. Where such data have been gathered, decreases in species richness have been detected. At the San Antonio cloud forest in the western Andes of Colombia.

Throughout the tropics, but perhaps especially in South America and Melanesia, the extent of local, regional, and full extinction of birds is undocumented for many species. This clearly justifies the call for more surveys. Many such surveys will be fascinating biologically but ultimately futile for preventing extinction as long as so much habitat continues to be lost. How personally frustrating it must be for scientists such as Gretton et al., who determined that only 20-30 pairs of Gurney's pitta still exist, yet territories of this colourful but secretive songbird are being lost each year to deforestation in Thailand.

The gap between scientific knowledge and conservation policy often is large. It may be safe to say that no single country on any continent has preserved enough habitat to secure the long-term survival of its current avifauna.

Islands

The relatively small land areas of islands result in small populations of organisms that tend to be more vulnerable to extinction than those on continents. Of 108 species of birds known to have become extinct since A.D. 1600, 97 lived on islands by their calculations, the probability of extinction during the past 4 centuries has been about 40 times greater on islands than continents. Although both 97 and 11 are underestimates, this ratio probably is more or less valid.

The plight of island birds did not begin, however, in A.D. 1600. Because of how incompletely we have sampled the zooarcliaeological record of island birds thus far, the known prehistoric extinctions are a small fraction of those that actually occurred. These losses seem to have been due to the same processes that still exterminate species on islands today: predation by humans and introduced species, normative pathogens, and habitat destruction. The losses of birds on oceanic islands consist of (1) extinction; (2) extirpation and (3) reduced population.

The last two categories are steps leading toward genuine extinction, which represents irreversible losses rather than the short-term fluctuations in populations near continental source areas that biogeographers often call "extinctions" when studying faunal turnover. Research in the Galapagos Islands. Hawaiian Islands, and Tonga has shown that natural rates of genuine extinction in island birds may be at least 2 orders of magnitude less than posthuman rates. Nearly all islands in Oceania were inhabited prehistorically by humans. Birds provided fat, protein, bones, and feathers for the human colonists, who also cleared forests, cultivated crops, and raised domesticated animals.

The prehistoric extinction of Pacific island birds is known from studying bones from archaeological sites. Micronesia is not as well studied as Polynesia, but seems to differ only in details of taxonomy and chronology. Although Melanesian islands also have lost a variety of seabirds and landbirds, a larger percentage of the indigenous avifauna survives on large Melanesian islands than on Polynesian or Micronesian islands, or on small Melanesian islands. While Pacific islands have a well-earned reputation as the source of much modern extinction of birds, most species of landbirds already were extinct when Captain Cook opened the region to European influence 220 years ago.

Lost were as many as 2,000 species of birds, dominated by flightless rails, but also including moas, petrels, prions, pelicans, ibises, herons, swans, geese, ducks, hawks, eagles, megapodes, kagus, aptornithids, sandpipers, gulls, pigeons, doves, parrots, barn-owls, strigid owls, owlet-nightjars, and many types of passerines. Assuming that about 9,600-9,700 species of birds exist today, the world avifauna would be about 20% richer in species had islands of the Pacific remained unoccupied by humans. In the Hawaiian Islands alone, at least 62 endemic species of birds have become extinct since the arrival of Polynesians nearly 2,000 years ago. Most of the extinct Hawaiian species were gone before the arrival of Europeans, including most species within the spectacular endemic radiations of cardueline finches and flightless geese, ducks, ibises, and rails.

In New Zealand, at least 44 endemic species of landbirds have become extinct in the past millennium, featuring many endemic species of moas, waterfowl, hawks, rails, and passerines. Outside of the Hawaiian Islands and New Zealand, prehistoric Polynesian birds are known from the Marquesas Islands, Society Islands, Cook Islands, Henderson Island, Easter Island, Samoa, Tonga, and Polynesian outliers in Melanesia. In these island groups, rails, pigeons, doves, and parrots have undergone the most extinction.

Rails have lost more species than any other family. Most island rails, in the Pacific and elsewhere, were flightless forest species rather than volant wetland or grassland species. Any Pacific island with a relatively thorough prehistoric record of birds has yielded bones of one or more endemic species of flightless or nearly flightless rails.

The three islands with the best fossil records each have produced two to four endemic species of flightless rails; the records from even these islands are incomplete. Two species of rails occur among only seven bones of landbirds known from remote Easter Island. The only surviving flightless rails in the tropical Pacific are three species of *Gallirallus*, *Nesoclopeus woodfordi* of the Solomon Islands, and *Porzana atra* of Henderson Island. All except the last species is endangered. Just think of the possibilities for studying comparative systematics, evolution, biogeography, ecology, and behaviour if hundreds or thousands of species of flightless rails still were alive.

Species nesting on or within the ground have been lost to predation from nonnative mammals and erosion of topsoil associated with deforestation. The loss of native birds is more complete on remote Easter Island than on any other island of its size in Oceania. Although Easter Island was forested at first human contact about 1500 years ago, deforestation was virtually complete by about 550 years ago. Small samples of bird bones, associated with Polynesian artifacts 600-900 years old, show that Easter island once sustained at least 22 species of seabirds, including 12 tubenoses. Only 1 of the 22 species of seabirds still nests on Easter Island itself, while 7 of them occur on one or two of its offshore islets. When fully known, the prehistoric seabird fauna of Easter Island probably will exceed 30 species, more than any other Polynesian island.

Bones also provide the only evidence that indigenous landbirds once lived on Easter Island, these being 6 extinct endemic species. On Ua Huka, Marquesas Islands, the Hane archaeological site has yielded thousands of bird bones. Eight of the 20 species of seabirds from the Hane site no longer nest on Ua Huka or its tiny offshore islets. Most of the 12 other seabirds nest only on the islets, not on Ua Huka itself.

The most common landbirds from Hane are 6 species of pigeons and doves, only 1 of which survives on Ua Huka. One subspecies of seabird and 8 species of landbirds from Hane are extinct. The prehistoric exploitation of birds can be evaluated precisely at Tangatatau Rockshelter (site MAN-44), Mangaia, Cook islands. The age of cultural deposits at MAN-44 ranges from about 1,000 years old to 200 years old. The eight extinct and five extirpated species of landbirds from MAN-44 are rails, sandpipers, pigeons, doves, and parrots whose bones dominate zones 1-4 and decline sharply by zone 5.

Most species of indigenous landbirds are not recorded above zones 5-7, which are 700 to 500 years old. The most common flightless rail, Porzana rua, is last recorded in zone 8. Two species of doves are the only extinct/extirpated landbirds recorded above zone 8. Each of the four surviving landbirds from MAN-44 tolerates forest clearance. The bristle-thighed curlew, a rare shorebird that nests in Alaska, has been killed and eaten for centuries on its Pacific island wintering areas, where about 50% of the adults become flightless during wing molt.Because of its depauperate modern avifauna, the Kingdom of Tonga did not qualify as an EBA in Bibby *et al.*

Bones from caves on 'Eua indicate, however, that at least 27 species of landbirds lived on this Tongan island before the arrival of people about 3,000 years ago. Only 6 of these same species still survive on 'Eua. If they were still alive, most species of 'Eua's extinct forest birds probably would qualify as RRS under the criteria of Bibby et al., thereby changing considerably the international conservation priority of Tongan birds. The Euan avifauna has been depleted irreversibly. As sampled thus far, the prehuman landbirds consisted of 27 forest and no nonforest species, compared to 9 forest and 4 nonforest species today. Forest frugivores/ granivores have declined from 12 to 4, nectarivores from 4 to 1, omnivores from 3 to 0, insectivores from 6 to 3, and predators from 2 to 1. These losses are more complete for ground-dwelling than midlevel/understory or canopy species, which might be expected with predation from humans, rats, dogs, and pigs.

7

Taxonomic Diversity

A project to study the taxonomic diversity of a smaller subgroup within the pyraloids is possible and will advance our knowledge and communication about this economically important group of moths. The Pyraloidea, like every group of organisms, has had many unique biological and historical influences that affect its present taxonomic status. This is a specific report on the initiation, management strategies, and progress of a study on the taxonomic diversity of the Pyraloidea of Costa Rica. Snout moths are primarily tropical in distribution, but occur worldwide, including most oceanic islands. They are known from low and middle elevations in the tropics to the high arctic.

While many pyraloids may have broad continental and even worldwide distributions, there are many subgroups that are endemic to islands and continental areas and that contain examples of extreme disjunction. Neotropical pyraloids comprise about 20% of the world fauna. The largest subfamily is the Pyraustinae, and to date 363 species have been delineated in Costa Rica. There are a few small subfamilies found only within the Western Hemisphere, such as the Midilinae and Linostinae.

Pyraloid caterpillars have diverse habits. They consume dried or decaying plant or animal matter, wax in bee and wasp nests, and living plants. Some are known to be inquilines in ant nests, predators of scale insects, and aquatic scavengers in flowing water. The Nymphulinae, with over 700 species, is the only taxon in Lepidoptera with truly aquatic caterpillars. Snout moth caterpillars are associated with nonvascular and vascular plants, feeding on both gymnosperms and angiosperms. They demonstrate plant specificity at different taxonomic levels. Because of the close association of the immature stages with plants, moths in general have shown a rapid response to environmental disturbance.

Changes in plant communities can be monitored easily and studied efficiently because adult moths can be sampled in a qualitative and quantitative manner with lights, such as mercury-vapour lamps placed in front of a white sheet or a Robinson-pattern light trap. Although the Pyraloidea have been included in many qualitative diversity studies, they have been included only recently in quantitative studies. Barlow and Woiwod utilized pyraloids to compare the seasonal and overall diversity of two lowland sites in Sulawesi.

Robinson and Tuck compared the diversity of primitive moths, specifically delineating subfamilies in the Pyraloidea, in three vegetational areas of Borneo. Pyraloids can be used in diversity studies because many are large enough to be sorted visually to species and are readily identifiable using collections such as the Natural History Museum in London or the National Museum of Natural History in Washington, D.C. The Pyraloidea consist of two families.

The Pyralidae are the smaller group, with only about 5,000 species and 5 subfamilies, and the Crambidae are the larger group, with over 11,000 species and 15 subfamilies. Some groups of pyraloids are brightly coloured, large in individual size, small in numbers of species, and have been well collected. The Midilinae and *Cliniodes* Guenee are over 2.5 cm in wing length; have distinctive wing patterns and colours, such as white, orange, and yellow; and have had a history of taxonomic work and illustration. In contrast, many pyraloid groups are not visually pleasing and are small in individual size and large in numbers of species.

For example, the Glaphyriinae and Scopariinae are small in size and have a homogeneous colour pattern; the Pyralinae in Africa and Nymphulinae in southeast Asia are speciose in poorly collected areas. Nevertheless, some pyraloid groups have been treated recently in revisionary or phylogenetic studies, such as the Phycitinae and Epipaschiinae, or in geographic studies, such as the Phycitinae, Pyraustinae and related subfamilies.

Snout mouth and agricultural diversity

Many pyraloids are pests to a wide array of crops and stored products. Some pyraloids rank among the most destructive pests to graminaceous crops such as corn, sugarcane, and rice. Table 1 illustrates the diversity of commodities affected by pyraloids and lists species that have been targets of concentrated worldwide efforts of control with parasites and predators. Although not listed, each genus may include closely related species that are also pests. Although the impact of snout moths on the human condition is based on their adverse effect, it is these same characteristics that have prompted testing of species of pyraloids for the biocontrol of noxious weeds.

The most striking success story is *Cactoblastis cactorurn* (Berg) (Phycitinae). It was introduced into many countries for the control of *Opuntia* Mill, the prickly pear cactus. In 1787, Opuntia was introduced into Australia and, by 1925, it had infested more than 24,000,000 hectares. But by 1934, after introduction of *C. cactorum*, 90% of the *Opuntia* stands were eliminated and the land restored for agricultural use. Some pyraloids, including species in the Nymphulinae with aquatic caterpillars, have been investigated for the control of aquatic weeds such as water hyacinth, hydrilla, and alligatorweed.

The study of the taxonomic diversity of pest species and their relatives is important for biocontrol studies, primarily for the importation of effective parasites and predators, and in the case of noxious weeds, for the diversity of insects feeding on the plants for experi-mentation as control agents. As a result of their notoriety as pests and control agents, some pyraloids have been well studied taxonomically, for example in the Schoenobiinae, Crambinae, and Pyraustinae. In addition there have been studies of snout moth guilds on economically important host plants such as legumes, cacti, and aquatic plants.

Biodiversity in costa rica

A combination of historical and ongoing events allowed a taxonomic diversity project in Central America, and specifically in Costa Rica: creation of the Instituto Nacional de Biodiversidad (INBio) in Costa Rica, organization of the largest neotropical pyraloid collection in the world at the National Museum of Natural History, a group of pyraloid taxonomists interested in biodiversity research, and publication of a checklist of snout moths for the entire neotropical region by Munroe *et al.*

Two prerequisites for a biodiversity study of an area are thorough collections from the study area and detailed taxonomic research. Large collections with long series of specimens from the entire range of the species are absolutely necessary to study variation and permit accurate identification of the species. Taxonomic research demands knowing the names and identities of all the species that have been described of a, particular group from the area. This list, in turn, allows the taxonomist to solve basic problems such as synonymy and recognition of new species.

The decision to undertake a study of the diversity of neotropical snout moths with Costa Rica as a taxonomic center was fueled in part by the creation of INBio, because it fulfills one of two important criteria, that is, thorough collections in the area of study. The Institute was created to document the country's flora and fauna in a creative and organized way. INBio through the efforts of parataxonomists, has extensive collections of Pyraloidea from national parks and conservation areas in Costa Rica.

Parataxonomists are Costa Ricans trained by INBio to collect and prepare biological organisms near their homes, and they are providing a thorough survey of Costa Rica in a relatively short period of time. The parataxonomists deposit the material at INBio where moth specimens are first labelled: a traditional insect label and a bar-code label. The bar-code label represents information about locality, collector, date of collection, and a unique number for each specimen.

The label is read by a bar-code laser light and placed into a computer database which is then made available to taxonomists working on the material. The inventory manager of pyraloids at INBio sorts, organizes, sends material to taxonomists, and enters information into a database. Another important consideration with respect to collections is the Pyraloidea collection at the National Museum of Natural History.

Although it is the largest neotropical pyraloid collection in the world, in 1989 over two-thirds was unsorted material and the remaining one-third was based on nineteenth century taxonomy. Over the past 4 years, a massive sorting and reorganization of the snout moth collection has made the material more accessible for taxonomic research and general retrieval of data.

Munroe estimated the total snout moth fauna in Costa Rica to be 2,000 species. To date, over 600 species from Costa Rica have been identified. Because of the size of the group, it was necessary to make decisions concerning the type of research required to answer specific taxonomic problems in each of the 17 subfamilies occurring in Costa Rica. Some subfamilies or certain genera with moths that are individually, very small, and where very few species have been described from Central America were targeted for preliminary taxonomic research papers primarily to describe the fauna. This approach has been found to be fruitful.

For example, the Glaphyriinae of Costa Rica comprise about 19% of the total neotropical glaphyriine fauna. Approximately 27% of the glaphyriine species in Costa Rica a new; specifically, 22 species had been described previously and 8 are new species. While this proportion of new species may seem high, it was expected because the moths are small and had not been well collected. Although the work continues in all subfamilies, the Phycitinae and Crambinae are the only two speciose groups that are still in the preliminary stage of identification.

The Pyraustinae, one of the largest subfamilies with about 1,500 species in the Neotropics and over 350 in Costa Rica, are still a major area of research because they have a number of genera, such as *Omiodes* Guenee, which are riddled with problems of synonymy and species complexes. The goal of this project is to produce a handbook for the identification of pyraloid moths in Costa Rica. But to solve many of the taxonomic problems, such as synonymies, complexes, and discovery and description of new species, research papers prior to the handbook would be more efficient, appropriate, and useful.

The first step in a study of taxonomic diversity is to know the names of all the species that have been described from the area. In the field of taxonomy, this is known as a checklist or a catalog derived from the literature. To be taxonomically informative and useful for future systematic studies, a list of the species should be based on a study with worldwide scope. This is especially true for snout moths because many genera are pantropical.

Fortunately, such a study has been conducted by Munroe *et al.*, for the neotropical region, based on worldwide generic concepts; this study provides the basis for the study of a geographical subset within the neotropical region. A checklist is just that, a list of names. What a taxonomist really needs is the identity, or morphological criteria, for assigning a name to a particular species of snout moth.

Technically, this and many other taxonomic problems are solved by studying and dissecting type specimens, a single specimen on which a species name and its description is based. Dissection and comparison of Costa Rican species to type specimens help to maintain nomenclatural stability and enhance worldwide communication. Type specimens of species usually are deposited by the taxonomist who described them in collections of museums, such as the National Museum of Natural History in Washington, D.C., and the Natural History Museum in London.

Although both are large repositories for type specimens of the Pyraloidea, type specimens also can be found in many other museums throughout the world. A checklist also provides the basis for efficiently solving the common problems of classification and indirectly aids in the discovery and description of new species. In any group that has not been revised recently, there are usually a large number of synonyms, i.e., many names for the same species described from different countries over the years by different people.

For example, the same species could have been described by Moschler from Puerto Rico, by Dyar from Panama, by Schaus from Costa Rica, by Druce from Mexico, by Amsel from Venezuela, and by Dognin from Ecuador. It was common practice in the last century to describe new species from different countries without communicating with museums, other workers, or the literature to determine if the species previously had been described. Another problem is that of a species complex or one name for what is really many species. This occurs usually because the moths externally look the same and no one has studied them in detail over their entire distribution.

Discovery of new species is an automatic by-product of studying the entire fauna taxonomically, not regionally. The actual description of all new species is not necessary or feasible at this point, and the decision of which new species will be described is based on a biological need to unravel complexes of species and to provide information about groups or taxa that have been historically underrepresented in the taxonomic literature.

Benefits of research on pyraloidean biodiversity

The ultimate benefit of the Costa Rican Project is

transfer of information within and from the field of taxonomy to all levels of society via several different vehicles such as literature and education.

The project will summarize information held in libraries, collections, and by taxonomists and place this knowledge at the fingertips of ecologists, conservationists, quarantine officers, and farmers in Central America. It will also be the first step to "taxonomic self-sufficiency" of pyraloid moths for Central American countries. This self-sufficiency will be accomplished by producing accurate scientific names and detailed characters that allow individuals to identify and communicate specifically and comparatively about the diversity of organisms on a global basis. The most tangible product that will reach a diverse audience is a handbook for the identification of pyraloid moths in Costa Rica with photographs, the correct scientific name, distribution, and taxonomic and biological notes on each species.

Analysis of the distribution of Costa Rican species throughout the Western Hemisphere also will be included and will form the foundation for other kinds of studies in other areas of Central America. Transfer of information via education has occurred at all levels of this project. In 1990, a group of parataxonomists participated in a month-long training course on how to collect, identify, and prepare pyraloids.

8

Living Resources

Protection and conservation of living species including domestic species, must, therefore, become an important issue for public policy and public pressure. In recent years, the government has begun to take conservation of wildlife seriously. But the manner in which these conservation programmes are being conceived and implemented is creating more tension than peace and harmony. Whereas conservation is on issue of crucial importance, it cannot be undertaken without the involvement of the people, especially those directly affected.

Nature conservation programmes cannot be promoted in the same ham-handed way as those programmes which have destroyed nature. Plant and animal life, as it has evolved on this earth, has produced a bewildering diversity. There are millions of species of plants and animals, but even within a species, there exists enormous diversity. For instance, there are an estimated 1,20,000 types of rice plants alone in the world and this is only a guess.

There can actually be many more. This massive diversity, or "gene pool", as scientists like to call it, is a common heritage of all human beings. It is also their insurance for future survival. Increasingly, as new varieties of plants and animals useful to human beings are bread, this massive gene bank provided by nature becomes important. Breeding is essentially an exercise in finding the desired gene from nature's vast genetic reservior and introducing into the genes normally used.

The Green Revolution, for instance is essentially based on a strategy of genetic reconstruction aimed at producing new varieties of cereal plants which can produce high yields. Scientists involved in this work, however, have also realised that the Green Revolution is something akin to riding a tiger; once you are on it, you have to stay on it to survive. To keep the Green Revolution going, breeders must produce new varieties every five to seven years as older high-yielding varieties succumb to pests and diseases, which themselves keep changing in intesity and character. If one variety covers too large an area for too long a time, it becomes an open invitation to disaster. A disease can sweep through a genetically uniform crop like wildfire leading to widespread crop failure and famine.

The brown rust infection, for instance, affected large areas in north India planted under the wheat variety kalyansona during 1973-74. Other than high yeilds, a high priority for both plant and animal breeders is the development of resistance to diseases, and screening available plant and animal varieties for these traits accounts for substantial amount of time and resources of breeders. Ironically using "superior" genes on a massive scale to increase production, as exemplified by the Green Revolution in food crops, blue revolution in fish production, white revolution in milk, and massive afforestation with high yielding monocultures, has posed one of the biggest threats to the world's genetic diversity in livestock, fish, crops and trees. In each of these areas, human beings are promoting a narrow genetic base—a few genes with highly desirable characteristics—to increase production.

In this process, farmers can replace thousands of wild varieties and primitive cultivated varieties overnight with one high yeilding variety. If the displaced varieties are not collected and properly stored, they could be lost forever. India has already lost thousands of traditional varieties of wheat since the beginning of Green Revolution. Simultaneously, environmental destruction caused by shifting cultivation, deforestation and dam building is destroying many wild varieties of even domesticated crop plants, trees and animals.

Water pollutions affecting fish stocks and probably even genetic diversity. Overgrazing is affecting grass and free populations. However, no gene in this diverse nature can be considered useless and, thus, deserving to be driven to extinction. India possesses an extremely rich diversity of plant and animal life. The Indian region is considered one of the 12 centres of genetic diversity in the world.

The total Indian flora consists of about 45,000 species of which about 30,000 are non-flowering plants like algae and fungi which are usually inconspicuous and small. The flowering trees and plants of India consists of about 15,000 species but this number should not be taken as the final figure as several areas in the country remain unexplored and underexplored. From a very small area, silent Valley in Kerala, covering less than 9000 hectares, the Botanical Survey of India recently discovered nine species new to science and a new genus of plants.

About 7000 endemic species are found in India, that is they do not occur any where else in the world.

Of these the Himalayas and the Khasi Hills in Meghalaya account for about 3000 endemic species, and the Deccan peninsula for about 200 species. The area of maximum diversity in India is northeastern India which is also the most threatened.

The second major area of genetic diversity is the lower region of the Western Ghats around the Nilgiris. Over 1500 of the rare flowering plants are subjected to a lesser or greater degree of threatened extinction. Just as India has benefitted from plant and animal genes from all over the world, India, too, has given to the rest of the world many valuable genes. Probably the most dramatic case is that of a single sample of wild rice, Oryza nivara collected by botanist S.D. Sharma from eastern U.P. in 1963. Grassy stunt virus is one of the four major diseases of rice plants.

During the early 1970s, grassy stunt epidemics destroyed more than 116,000 hectares of rice in Indonesia, India, Sri Lanka, Vietnam and the Philippines. At the International Rice Reserch Institute in Los Banos (Philippines), Gurdev Khush and K.C.Ling searched 6723 samples of cultivated and wild rice varieties looking for resistance to grassy stunt virus. Only one sample, that of Sharma's displayed the resistance they were looking for.

Today, the genes from this sample are used in resistant varieties grown on 30 million hectares in India, Nepal, Bangladesh, China and Southeast Asia. The disease has almost ceased to exist in Indonesia. These genes have helped push average production per hectare to six tonnes and in some cases to 10 tonnes. Two or three crops have become common and rice production overall has doubled. In an interconnected world, Sharma's rice sample has helped India to repay a debt. Indonesian wild genes have helped India to set up a sugarcane industry.

The limiting factor to commercial sugarcane production in India used to be the red rot diseases. The resistance was acquired from the wild Indonesian cane Saccharum spontaneum, Indonesian S. spontaneum genes are now present in important commercial varieties of sugarcane used in such top sugar-producing countries as Cuba, Mexico, USA, Argentina, Brazil, Columbia, China, India, Indonesia, Pakistan, Philippines, Thailand and Australia. Today equally widespread are S. spontaneum clones from India, which are sources of resistance to gummoss and root rot.

The list of valuable genes discovered in India is long. Large areas of the world, for instance, are covered by unproductive saline soils. Tolerance to salinity in rice has been tound in several traditional varieties from southern India and Sri Lanka. A wild melon from India has provided resistance to a form of mildew and has saved the melon crop in California, USA. Among animals, too, many Indian breeds are known all over the world. The home tract of the famous sahiwal cattle is in the Sahiwal district of Pakistan. At the time of partition, only a few Sahiwal herds existed in India.

A few more animals were brought by migrating Sikh and Hindu families. Today, the Sahiwal population in India consists of about to 1000-1100 breeding females. The breed is being used for crossbreeding in many countries. Sahiwal semen is being produced on a large scale at a farm in Queensland (Australia) for export to New Zealand, which, in turn, after contract matings by use of Sahiwal semen on Jersey and Friesian cows, exports crossbred heifers to Southeast Asian countries.

The famous Murrah, Nili-Ravi and Surtibuffaloes are being used to upgrade buffaloes in many countries of the world. In Vietnam, a consignment of

Murrah buffaloes is widely regarded as the most valuable gift from friendly India. The Jagarabadi has been successfully introduced into lands in which buffaloes did not exist untill recent times, for example, brazil and Trinidad.

The Jamnapari goat breed of the Chambal ravines in Etah district(UP) has been used for improvement of native breeds in several countries. The famous English Anglo-Nubian breed, which has been extensively used for crossbreeding, is itself based on a cross of the Jamnapari. Even germplasm of economically important insect needs to be preserved. The two insect livestock groups of major world importance are honey bees and silkworms. There are many strains of honey bee, and there is great variation within the strains for characteristics such as honey production and disease resistance.

Though efforts to collect wild bee germplasm seem to be a long way off because of various technical reasons, the need is being increasingly felt. Work on breeding has already been done with the mulberry silkworm. In India, wild forms of the tasar silkworms (Antheraea species) are being used for improvement. Antheraea roylei from India and A pernyi from China have been crossed to produce the hybrid A, proylei. Silk from the hybrid is claimed to be the finest ever produced in India and the new species also out yields its parents by almost 170 percent in weight of silk and 94 percent in filament length.

The hybrid is expected to play a major part in eforts to expand the tasar silk industry and provide rural employment. In a recent lecture on genetic conservation, Dr. M.S. Swaminathan, one of India's leading agricultural scientists and Director-General of the International Rice Research Institute, drew attention to the need to conserve even micro-organisms. There has been an enormous growth of interest in biological nitrogen fixing organisms because of the ill-effects of chemical fertilisers on the environment. Among the major sources of biofertilisation are blue green algae and azolla. Azolla germplasm from all over the world is already being collected at the International Rice Research Institute.

Azolla is an aquatic fern capable of fixing atmospheric nitrogen through a symbiotic blue green algae living in the cavities of its leaves, it has long been used in China and Vietnam as green manure in rice fields and its use is growing in other Southeast Asia countries flow. Blue green algae (BGA) is being extensively promoted in India as a cheap source for fixing nitrogen in paddy fields. Unfortunately, there is no organised effort to collect BGA for genetic conservation purpose. Strains recorded during ecological studies are not kept by the scientists because maintenance of a culture collection is time consuming and expensive.

Scientists have achieved considerable success in selecting strains of Rhizobia nitrogen fixing bacteria which live in the roots of leguminous plants, and give them their nitrogen fixing abilities. This success indicates the possibility that superior strains may exist in other nitrogen fixing organisms, too. Today, the world's main food supply is determined by only 20 crops specis, and only eight forest species provide most of the world's wood and timber. Similarly, about 20 animal species provide most of the world's milk, wool, hide and draught power.

A much large number of plants to provide food, such as oil, vegetable, fruits and fodder, and meet other needs, such as spices, beverages, medicines, ornamentals and industrial raw materials. However, all these plants and animals constitute only a fraction of all the plant and animal species in the world. It is impossible to guess what this natural wealth will yield in the future. But the potential can be seen from the vast variety of medicines alone that plants have given to human kind. The Ayurvedic system of India gave to the world reserpine, a tranquiliser and a treatment for hypertension, anxiety and schizophrenia. Reserpine is a therapeutic agent extracted from a herb, Rauwolfia serpentina, which is now almost impossible to find in forest. In India, at least 2500 plants are used for medicinal purposes.

The roots of the berberry shrub found in the Himalayan foothills are an important source of berberin, a compound used widely in tropical Asia and Japan against diarrhoea. A single species, Strychnos nuxvomica, indicates the commercial value of drug producing plants. A source of the compounds strychnine and brucine, species earns India over Rs.50 lakh yearly in exports.

India's total exports of plant-based chemical products, mainly medicines and pharmaceuticals, are worth over Rs.55 crore yearly, a trade 'which increased five-fold during the 1870's. It is distressing that overexploited medicinal plants face probably the greatest threat of extinction in India today. From an ethical point of view too, there are strong arguments against human beings driving any life form into extinction. But protecting the wonderful gene bank provided by nature is an issue on which both utilitarian and ethical perspectives coincide. India is known especially for its rich genetic wealth of bamboo, fruit trees (mango, banana and citrus), vegetables like eggplant, akra and cucurbits (melons and gourds), grain legmes (mung beans urad or black gram and cowpeas), rice, pepper, cardamom, ginger, turmeric, sugarcane, jute, and various medicinal and aromatic plants.

There is also a rich diversity in wheat, maize, millets, oilseeds and various other fruits, spices, vegetables, fodder and plantation crops. Plant species from India have not only influenced agricultural development in the region but also in many other parts of the world. Plants were taken from India to the ancient civilization of the Tigris, Euphrates and Nile river basins.

Many Arab traders took plant materials to Meditarranean countries in the 8th-10th centuries AD. Buddhists took several Indian cultivated plants to Southeast Asian Countries. It is estimated that about 250 India species have wild plants of agricultural and horticultural importance and of these about 60 are rare or endemic. Though many wild varieties have neen collected and used in breeding programmes, only a fraction of the job has been done. India is also very rich in its natural wealth of forage legumes and grasses. About 2000 such species occur, of which tow-thirds are grasses. The northeastern Himalayas are a veritable mine of valuable genes in many crops and fruit trees. Unfortunately because of shifting cultivating and widespread habitat destruction, this genetic wealth faces the threat or extinction.

Dr. Bhag Singh of the National Bureau of Plant Genetic Resources (NBPGR), for instance, discovered *citrus indica*, an endangered species of citrus, in the Tura range of the Garo Hills, Meghalaya. This species is the most primitive of citrus species and perhaps the progenitor of cultivated materials. Citrus fruits are native to South Asia and the northeastern Himalayas hold a notably important position in the world with regard to citrus wealth.

The Khasi and Garo tribes grow citrus trees in and around the vicinity of their houses. A major crop in India and Asia, rice has attracted the maximum interest amongst indian scientists interested in plant genetic resources. The areas with maximum diversity in India are the northeastern tracts and the hilly regions of Orissa, Bihar, Madhya Pradesh and Andhra Pradesh, which are mostly inhabited by tribal population. Various local rice varieties have also been found in the malabar region and in the cold Himalayan tracts. A national survey estimated that although about 50,000 local rice varieties are still in cultivation, they are being rapidly lost, particularly in the irrigated tracts, where new high yielding varieties have been adopted, growing deforestation poses a serious threat to many wild varieties of rice.

Tropical rain forest habitats have been important sources of disease and pest resistant varieties of rice. The Silent Valley forest reserve area in Kerala, which was to have been inundated with the construction of a hydroelectric project, has been the Northeast are another invaluable source of rice germplasm in the form of the well-known Assam Rice Collection (ARC). Together with the threatened cultivated varieties, these endangered natural habitats should be a priority area for immediate collection of wild varieties.

In a germplasm conservation programme, the wild species are as important as, or more important than, local cultivated varieties. A total of nearly 15,200 local rice varieties are being maintained at the national rice gene bank at the Central Rice Reserach Institute (CRRI) at Cuttack, and an estimated 35,000 local varieties are held as working collections at 40 rice reserch stations. Thus there is a total collection of 50,000 cultivars (Cultivated varieties), Assuming 50 percent of the collection to be duplicates, the national collection is now estimated at 25,000 with scope for further collection of another 25,000 varieties. The genus Oryza (rice) consist of about 24 species, out of which only two are cultivated: O sativa (Asian rice) and O glaberrima (African rice). Collection of wild varieties has logged behind those of cultivated varieties.

The group O, rufipogon, O, nivara and O, meridonalis, is the most closely related to the widely cultivated O. sativa rices and represents the progenitors or the cultivated rices. Although as genetic resources, their importance is paramount, unfortunately, they are fast disappearing. This group occurs in one of the densely populated areas of the world, where the habitat is being reclaimed for cultivation of crops, construction of roads and buildings.

The cultivation of O.sativa rices is coming closer to the population of these wild species which are hereby losing their genetic identity by being "swamped" by the genes of O.sativa. A time bound collection programme for the wild rices in this group is thus urgently needed. The group O.Collina, O.officinalis, O.minuta and O.malampuzhaensis are Asian rice species which occur in forests in partial shade, by the side running streams or swamps. With deforestation, this group is also endangered.

The present collection of O.officinalis from South Asia is not representative of their natural distribution. It is important to collect these species before they disappear. In India, the CRRI is aided by 10 base centres at Simla, Almora, Jodhpur, Ranchi, Shillong, Cuttack, Hyderabad, Akola, Trichur and New Delhi. These centres collect not just germplasm for rice but also for awide range of other crops. Each centre has been allocated a state or region in which it actively collects. In future collections of wild rice cultivated varieties,upland rainfed rices, flood-tolerance, pest and disease resistance, and aromatic and qualitycharacters will be emphasised.

Collecting these wild species and peripheral populations under severe pressure may require specialised collectors and the creation of "gene sanctuaries" to protect threatened habitants.

Accordign to Dr. M.S.Swaminathan, the collection of germplasm should not only involve professional scientists with expert knowledge of rice but also farmers, the general public and the defence services. In China, for example, each provincial academy mobilises farmers to send in seeds of traditional varieties.

Although some samples will duplicate, this system lessens the work of the scientists and lets them concentrate on classifying the material and identifying duplicates and distinct types. As some of the high priority germplasm collection areas and end angered habitats are either sensitive border zones or inaccessible regions, defence personnel could also play an important role in collecting germplasm.

Amongst the most vulnerable plant species in India are overexploited medicinal plants, ornamental plants, like orchids and botanical curiosities like pitcher plants. Some 1250 species of orchids occur in India, of which 300 are found in Meghalaya alone. About 20 species of orchids are becoming scarce in Meghalaya. Scientists from the Botanical Survey of India (BSI) recently made intensive studies on a group of beautiful orchids called Coelogyne, but failed to see even one plant of three species, namely, Calbolutea, C.assamica and C.treutleri in living condition.

They have probably become extinct. Even their preserved specimens are not available in any herbarium. The Government of Meghalaya has recently established a "pitcher plant sanctuary" in the Khasi Hills. The Indian Orchidarium of Shillong has brought approximately 350 indigenous orchid speices into cultivation for conservation purposes. But much more needs to be done in this connection. For example, arhododendron sanctuary is urgently needed in Sikkim.

The loss of valuable plants is definitely a matter of great concern, though new attempts have been made as yet to collect their germplasm. Amongst the endangered medicinal plants are three orchids. Paphiopedilum druryi a native to kerala, which is endangered or extinct in the wild due to forest fires and excessive collecting; Dendrobium pauctiflorum found in West Bengal and Sikkim in areas open to tree felling which is also endangered, possibly extinct; and Diplomerial birsuta, restricted to a tiny population in West Bengal and is vulnerable to landsliders.

A decrease in its number has been noted by botanists. The numbers of another orchid, Dendrobium nobile are seriously declining. This orchid occurs in the Himalayan regions of India and China and is the source of dendrobine which is still exported from China by the ton in a dehyarated from. Several other medicinal plants are being seriosly threatended because of over exploitation.

A corus calamus, known as vacha in India, is used in 51 different drug preparations. It used to be common in Manipur and the Nega Hills, but today it is so scarce that it has to be imported from overseas, Rauwolfia serpentina, the source of resperpine, known to Ayurveda for centuries, was once common and abundant in forest areas all over peninsular India. But after scientists isolated the alkaloid reserpine in 1952, a ruthless search was started all over the country, which came to a halt only when the plant almost disappeared from forest areas. Today it is extremely rare in nature and is mainly found in cultivated stands.

Variations within the species may have been lost already. Two other species that are extensively used in Ayurveda and are today endangered are Saussurea lappa, the source of kuth roots, and Dioscorea deltoidea, the source of steroidal hormone. Today these plants are listed in the Convention on International Trade in Endangered Species and cannot be exported without an export permit. Dioscorea deltoidea, much sought after by private agencies and pharmaceutical firms, has been continuously collected in India except perhaps in the inaccessible areas of the Himalayas.

The roots yield cortisone, a steroidal hormone used in treating rheumatic diseases and ophthalmic disorders. The need to preserve the medicinal plants of India is clearly vital and urgent. Maintenance of botanical/herbal gardens will not be a solution. Efforts will have to be made to conserve the germplasm of each important species of medicinal plants.

The National Bureau of Plant Genetic Resources (NBPGR) has only made a beginning by collecting Opium Poppy, Syllium, Senna and Rauwolfia serpentina. The NBPGR has a major role to play in the protection of the country's vital genetic treasure. But the Indian council of Agricultural Research (ICAR), which supervises NBPGR, will have to be more effective in energising this organisation that it has been so far. Germplasm conservation is a complex task. Clear priorities have to be set in collecting germplasm.

There should be an understanding of how existing collections or duplicates should be weeded out. Collected seeds should be stored carefully. They must be properly characterised for their attributes and then made available to breeders for use. Finally, there should be adequate facilities for multiplication of seeds while duplicate samples should be kept as an insurance. Although the NBPGR was set up in 1976, it has still not built a proper facility for collected germplasm. Funds were allocated for the construction of storage facility in 1978 but it took three years just to choose the site. It has taken a further three years to finalise architectural details, with central public works department officials anxious to visit Europe to look at existing storage systems. There is still no firm date by which the storage facility will be built. In such circumstances seeds have to be sown every year and recollected.

One can only guess how well the specimens have been kept. It is quite possible that many samples have been lost or are mixed up. The state of the rice collection at the Central Rice Research Institute should also be a matter of public concern CRRI too does not have a long-term storage facility. In the humid climate of Orissa, the viability of rice seeds is generally not more than six months. Consequently, many scientists lack confidence in NBPGR and prefer to keep their collections with themselves.

The most extensive collection of rice from the Chattisgarh region was made by Dr. R.H. Richharla. Beginning in 1971, while workings as the agricultural adviser to the government of MP, Richria collected, with single minded devotion, some 18,000 samples of rice. Though many of these may be duplicates, Richaria's vast collection become the nucleus for the breeding activities of the MP Rice Research institute (MPRRI),which began in 1976 and was later merged with J.N. Agricultural University.

Richaria, in fact, argued that it is the higher-yielding among these indegenous varieties which need few inputs, many indegenous varieties will outyield the so-called high-yielding imported varieties. The work of collecting rice varieties in Chattisgarh has almost come to a halt since MPRRI was close down in 1978 allegedly under World bank pressure Doubts have been expressed about the care with which Richharia's collection has been kept.

India has a rich diversity of animal life which is threatened by a number of factors. Even as scientists discover new breeds like the Siri cattle in Sikkim, and although many species are nowhere near extinction, they are undergoing a process of genetic erosion. Urban centres attract cows and buffaloes with the highest milk yield from the rural areas. But from there they go to the slaughter-houses, after completion of location.

Thus, high producing animals (genotypes) are being regularly removed from the breeding tracts, leaving inferior genotypes for breeding. P.N. Bhat, one of India's leading animal geneticists, believes that the genetic quality of the famous Murrah breed has deteriorated. The maximum recorded milk yield in a location for a Murrah buffalo was obtained in 1932 with one female yielding about 4000 kg. In a lactation of 310 days, it is difficult now to find buffaloes recording better that 2500 kg per lactation. It has been estimated that between 1921 and 1979 the loss in average lactational yield of good producers due to the genetic erosion caused by slaughter-houses and other factors has been of the order of 1000kg. India has 26 breeds of cattle and seven of buffaloes, together with a number of nondescripts, which account for about 75 per cent of the total cattle population.

The improved and specialised breeds are generally found in the northwestern and western dry areas. Indiscdriminate crossbreeding with exotic breeds poses a serious threat to indigenous breeds. Already crossbreeding with temperate breeds has made in roads into good drought breeds. Bhat argues that the need for crossbreeding in their breeding tracts needs to be evaluated and stopped if erosion is large, so that the local breed may not be last. Some of the finest milk breeds are facing constant erosion, says Bhat, both because of indiscriminate crossbreeding and neglect. There have been no practical breeding programmes for genetic improvement of the indigenous breeds. These breeds are not in as good a genetic condition as they were a few centuries ago. Bhat himself believes that livestock breeds which can sustain and produce on low agricultural inputs should be identified and selected in animal rescarch programmes in tropical countries.

There is little point in introducing animals which need high inputs to produce efficiently in conditions of poverty. Apart from the humped cattle, there are in the northestern Himalayas a number of wild, semi-wild and domesticated humpless cattle. Economically the most important is the mithum (Bos frontalis), raised primarily for meat, sacrifice and bride price. The mithun is believed to have arisen from the cross of gaur, the wild cattle of the indo-Malaysion region, with domestic cows.

The 1972 census revealed 121,000 mithuns in Arunachal pradesh, Nagaland, Tripura and Manipur. The mithun lives mainly in semi-wild conditions at an altitude of 600-1600 metres. Among the lesser known cattle, but extremely important for Himalayan traders, is the yak, or grunting ox (Bos Poephagus graunieus Linn). In the wild it inhabits the coldest and most desolate tracts of the Himalayas, from the east to the west, with elevations ranging from 4000-6000 metres. Without the yak, travel and trade in these parts would be as impossible as without the proverbial camel in the desert. According to the 1972 census, there were 43,000 yaks in India.

The yak interbreeds with many types of cattle: bison banteng, goyal, zebu and European cattle, and has been so interbred for centuries. No interbreeding with buffaloes has been recorded so far in India. Yak bulls are mostly used for crossbreeding hill cows. The hybrids are hardier than the yak and are capable of bearing heavier loads, but are inferior in staying power. Hybrids with cattle are now twice the number of purebred yaks.

Since there has been no organised effort to preserve this useful animal, its population has declined by almost 25 percent over the last 10 years. Though sheep are very important in certain regions of India, little work has been undertaken to identify whether any valuable Indian sheep breeds face the danger of extinction. It may take as much as 25 years before a clear picture of the available genetic material in sheep emerges.

Scientists argue that evaluation of indigenous sheep breed for corssbreeding and their conservation should go hand in hand. It should be mandatory for all sheep development programmes undertaken by state governments to conserve nucleus flocks of breeds used for crossbreeding in their natural habitat. Like sheep, goats are also important to the Indian rural economy, and as one goat expert puts it, despite regular attempts by soil conservationsists to exterminate this animal, its population has been increasing.

The origin of Indian goat breeds is not clearly known. They are believed to have been derived from wild goats which inhabited the Asian mountains in antiquity, and to have been domesticated around 7th century BC much earlier than cattle. There are at present about 17 distinct goat breeds in India, of which several like the Jamnapaari are extremely useful. The dwarf goat breeds, such as Black Bengala, Banbari, maalabari and Assam Hill are famous for high prolificacy (multiple births) and early sexual maturity. The pashmine goats of Ladakh produce the finest quality of Pashmina fibre (cashmere) in the world. But though Pashmine goats support a major cottage industry in the Ksahmir valley, little is known about them. Research on goats is of recent origin and specific strains are now being developed for milk, meat and fibre, crossing of pashming goats with angoramight make possible a goat with mohair and poshmina. Gaddi, a hill goat, produces hair used for ropes. Crossbreeding of these goats with angoras yields mohair. India had only about seven million pigs in 1976 but they are an important source of meat espcially for enonomically disadvantaged groups like Harijans.

Traditionally indigenous pigs produced bristies and meat. In fact, meat was only considered a by-product, until the advent of nylon which replaced pig bristles. There has been no serious attempt to catalogue the various breeds or strains of domesticated pigs in India, and previously unknown varieties are some times found for example, P.N. Bhat recently located a number of strains of dwarf pigs in the north eastern region. Indigenus pigs can survive under extremely adverse conditions of malnutrition and health: They are reportedly more resistant to disease than exotic breeds.

Pigs are receiving greater attention under government programmes and indigenous pigs crossed with exotic breeds are becoming more popular particularly in central India. Their higher growth potential makes them attractive. but given the adverse circumstances in which they are maintained, the survivability of corssbreeds is low, and therefore their advantage is marginal. Because of extensive crossing with the Large White, the indigenous pig in the central region is becoming extinct. Indigenous breeds need to be properly documented and their performance tested so that proper breeding strategies can be decided. Besides domesticated species, there are several wild ones, such as the common Indian wild boar. The present day pigs have evolved as a result of gradual domestication of wild pigs. India is one of most important countries as for as camel husbandry is concerned its population of 1.1 million camels constitutes about seven percent of the world total. Camels are found mainly in Rajasthan, Haryana and Punjab, the Arabian one-humped camel being the only species.

There are three types of camels found in India: the desert camel, the riverine camel and the hill camel. Bikaneris the most important desert camel breed in India. Nearly 50 percent of all Indian camels belong to it. Riverine camels are found in districts of Uttar Pradesh and Punjab, where there is a good water supply. They carry heavy loads and are slow in movement with a speed of 3 km/hour. The desert camel can walk 190 km overnight at a speed of 15 km/ hour. The hill camel is more compact than the camels in the plains.

The National Commission on Agriculture (NCA) had expressed concern over the changes taking place in the traditional system of camel breeding practised over centuries in Rajasthan. Under the traditional system, camel breeding was undertaken by professional breeders under the patronage of princely rulers and large landlords, who maintained large herds of camel. Excellent camelherds were owned by Bikaner, grazing of camels with no tax levied on their grazing. Drinking water from all the wells in the state was allowed free for the animals. Good stud camels called "Mehias" were branded and let loose by individuals intrested in camel improvement.

These stud camel would move in herds in winter and during the offbreeding season they were free to graze anywhere, without fear of impoundment even if they damaged standing crop. The traditional camel breeders, with their keen observation, used to breed camels with a definite objective. The camel breeders in Rajasthan evolved distinct types of camels like the Bikaneri and Jaisalmeri.

Today, however, the conditions favouring breeding of camels are fast disappearing. Grazing lands have diminished at a rapid rate, having gone under the plough. As a result most camel breeders are now leading a pecuniary existence with small herds of camels in poor condition. Under these adverse conditions the fine art of breeding has disappeared, reduced to a mere multiplication of stock. "It may not be long before these camel breeders will go completely out of business", warned the NCA.

The NCA was not even sure whether there is a need to make a special endeavour to retain the distinct types of camels that were evolved in the past. "Breeding of animals for genetic improvement without ensuring adequate nutrition is meaningless", it arugued. The construction of roads in desert areas is making the animal itself redundant. In the past camels were prized most for their speed of travel and capacity to traverse long distances at a stretch in the desert. In increasing numbers, camels are today being employed for agricultural operations. They played an extremely important role in the construction of the Rajasthan Canal.

The survival of the animal, and its various types, is now dependent on whether adequate grazing lands will ever be found for it. India also has about 0.94 million horses and ponies, 0.075 million mules and 0.99 million donkeys. Uttar Pradesh has the greatest number of horses and mules, and Rajasthan donkeys. Horse breeding has been practised in India for a long time. There are six important breeds of Indian horses and ponies, which include the Kathiawan, one of India's best breeds which is found in Rajasthan and Gujarat: Marwari, which is closely associated with the great feats of bravery attributed to the rathor Kings of Rajputana: Bhutia, which is bred along the Tibetan border and sub-Himalayan tract from punjab to Darjeeling: Manipuri, which has been bred in Manipur and Assam for many centuries: the Spitipony of the Spito Valley: and the Zanskari horses of leh in Ladakh. Donkeys are of two types: The small Grey, which is commonly found in most parts of the country and the large white which is found in the Kutch. Unlike the camel, the donkey has never received much attention. there has been no organised donkey breeding in India even though it is an important work animal.

There are no textbooks or monographs devoted to the donkey, no research on its physiology, productivity or work ability, no selection experiments to increase its performance, even though next to the camel, it is the best farm animal adapted to a hot, dry environment. Chickens constitute over 90 percent of the total poultry population of India, followed by ducks, six per cent turkeys and geese, 2 per cent, India and the neighbouring countries have been referred to as the original home of the Red Jungle Fowl (*Gallus gallus*), from which present day domestic birds are believed to have descended.

The Aseel or Malay Fowlis reported to have given rise to all the breeds of poultry. There is evidence to show that these birds, some 2000 years ago, moved through the Middle East to Europe and gave rise to the present day European breeds. Eighteen types of indigenous breeds are on record. But in number, they are much smaller than exotic breeds. The former number about 50 million as compared to about 100 million of exotic breeds (including crosses). Several indigenous breeds are facing extinction. The original Aseel, a native of Andhra Pradesh, is a medium sized, round-bodied and well-built bird. However, large specimens have also been bred and developed by the promoters of cock-fighting. The bird is poor layer but is known for its delicious meat. With the ban on cock-fighting, the Aseel is losing much of its importance, and as a result pure specimens are becoming rare.

The Ghagus is a large sixed bird, which is usually found running with the nomads of Karnataka and Andhra Pradesh. It is a hardy breed and is supposed to have resistance to diseases and parasites. Due to neglect, it is gradually becoming extinct. Information on ducks and other birds is extremely limited, there appears to have been no evaluation of indigenous ducks. At a recent international meeting a scientist expressed concern over government plans to establish breeding farms for exotic Khaki Campbell ducks, both for intensive rearing of these ducks and for upgrading local ducks.

There is no specific information as yet that under normal husbandry conditions of limited feed supply and no-existent disease control, the exotics are better than the indigenous birds. Animal conservation is still a new area even for scientists. With the failure of several animal development programmes in which exotics have been introduced on a larger scale or crossed indiscriminately with local breeds, the importance of environmental adaptation using local breeds is being recognised.

Hence, the importance of conservation of indigenous breeds to support more appropriate breeding programmes is also acknowledged. Conservation of animal genes is neither easy nor cheap. It can be done either by maintaining a living herd or flock of animals that need to be conserved, by storage of frozen semen from a representative sample of males or by storage of frozen embryos. In India the need to conserve animal genetic resources is being recognised. Several institutes concerned with preserving animal genetic resources have been set up. They are the National Bureau of Animal Genetic Resources (NBAGR), a National Centre for studies on the Mithun in Arunachal Pradesh, a National Institute for Research on Goats in Mathura district, Uttar Pradesh and a breeding farm for camels of Bikaner. An enormous amount of work is needed to document and evaluate indigenous breeds of birds and animals and then based on this information, urgent action taken to conserve breeds. Though India has a rich variety of fish fauna only a few species have ever been cultivated.

The most important among these are the carps including Catla catla, Labeo rohlta, Cirrhinus mrigala, labeo calbasu and a few others. Several exotic species have also been introduced into India (the Chinese carps, tilapia, trouts, gourami, etc). The natural habitats of the cultivated species are the major rivers where they breed naturally. Under aquaculture, however, though they mature in ponds, they normally do not breed there.

Fish seed, therefore, has to be collected from the rivers for aquaculture. As it is being practised today, this will have a detrimental effect on fish and may leave a genetically deprecated stock in the natural habitat. Since fish species inhabit different ecological systems in geographically and climatically in isolated areas, it is possible some of these species could have evolved into definite strains having various advantageous characteristics. Unfortunately, no comprehensive study has been made so far even on these few cultivated fish. Intensive aquaculture through genetic selection and selective breeding result in the species becoming more inbred and with less genetic diversity.

Though aquaculture work in India is still limited, original diversity could be eroded in future. Thus to maintain the genetic diversity for present and future requirements, conservation of genetic resources in fish is essential.

Several dams and canals are being built to harness the country's water resources. They can have serious implications for fish genetic resources by permitting inter mixing and breeding of different strains and species which hitherto enjoyed isolation and preservation in different geographical and ecological boundaries. Transmittance of undersirable species to areas where they are not liked or which may compete for food space and breeding with local varieties; introduction of fish predators and various other organisms; introduction of fish parasites and dangerous pathogens to areas where they were not known before; and spreading of undesirable and harmful aquatic weeds are some of the more obvious possibilities. In the case of crop genetic resources, after exploration, varieties are conserved in a central pool.

Thus at any given time the required strain is available from a central place. But maintenance of fish at a central pool for a number of years can be hazardous as there can be complete loss of a precious variety through disease and other unfavourable factors. Furthermore, maintenance of certain strains over decades in ponds and repeated breeding will result in partial or complete domestication of the species. This is contrary to the idea of conserving genetic resource. The best way is to protect these varieties in their natural habitat.

This needs legislation to declare certain natural habitats protected and free from fishing; preventing pollution of the protected zone; preventing construction of dams and weirs in these protected zones or if dams and weirs are constructed, suitable and drainage system; protecting fish breeding areas in natural habitats; and, restricting large-scale exploitation of fish seed from riverine sources. The Indian Council of Agricultural Research (ICAR) has recently established a working party on conservation of genetic resources of fish to consider all aspects of this problem. But this constitutes little more than a beginning of the understanding of the problem. Large areas of forests are in imminent danger of disappearing.

Although there is little risk of entire species of trees becoming extinct, populations at the limit of the species range or in isolated block of forests are definitely vulnerable. Often these small populations have a high frequency of valuable genes-for drought resistance, for example-which could be used in future tree breeding programmes. Tree genetic resources have been neglected in the past. With a growing interest interest in forestry across the world now, it has becmoe essential to preserve natural wealth in trees.

South Korea, which has carried out the most extensive afforestation in Asia, has set up on entire research institute devoted to tree genetic resources and development of suitable trees. Tree populations are most threatened in arid and semi-arid areas where an intense pressure exists because of the demand for firewood. Developments in commercial forestry have put imense pressure on specific trees which have acquired an economic value.

The Dang forests in Gujarat, for instance, are a rich reservior of a number of plant species which were not considered of economic value about 20 years ago. Good examples of these species are kakad (Garuga pinnata), which is used for pulp and packing cases, tanch (Ougeninia oojeinesis) used for making agricultural implements, haldu (Adina cordifolia) and kalam (Mitragyna parvifolia) used for making agricultural implements, haldu (Adina cordifolia) and kalam (Mitragyna parvifolia) used for making bobbins and furniture, and modad (Lannea grandis) and kudi (Wrightia tinotoria) used for making toys. It is important that tree species in this area, as in many other forest areas of the country, be preserved. Equally, several village woodlots, with trees representing well-adapted genetic material, are being steadily depleted.

FAO wants to give priority to collecting the diversity in tree species suitable for firewood and fodder production in dry areas. Arunachal Pradesh is another area where the richness in tree resources is being steadily threatened. As two scientists of the Forest Research Institute of Dehradun put it: "In spite of the richness of the flora, the area, to this date, remains a terra incognita and without a comprehensive published flora". A paper published in "The Indian Forester" in 1980 listed abut 20 species facing danger of extinction or depletion.

This list includes Phyllostachys assamica, a bamboo used for making huts, Gymnociadus assamicus, a small tree which grows up to 20 metres high and whose fleshy pods are used for washing hair as a substitute for soap, and the spectacular Rhododendron dolhousiae which has cream colour flowers with purple bands, R. dahousiae is such a beautiful plant that it was awarded a first Class Certificate by the Royal Horticultural Society in the UK. It is now facing extinction owing to massive road building and developmental activities in Arunachal Pradesh. Tree genetic resources can be conserved either in situ or ex situ.

Conserving in situ, that is, preserving the entire ecosystem together with its trees is definitely more desirable but often extremely difficult and expensive. When effective in situ conservation not possible, trees have to be conserved in artificial stands established ex situ in their country of origin or elsewhere. Conservation of tree genetic resources has been practised in India through the system of "preservation plots". The first preservation plot was set up in the sal forests of Bihar in 1906.

Some 188 plots have been established since then to protect mainly commercial species like Shorea robusta (sal), Tectona grandis (teak), and various conifers. Most of these plots cover natural forest types and range from a few hectares to 4000 ha. Preservation plots in plantations are small. The coverage of these plots is also patchy. Many forest types like southern tropical forest, semi-evergreen forests, and tropical thorn forest, need particular attention.

A beginning is being made in the preservation of trees important for social forestry, Leguminous shrubs, which can enrich soil fertility through biological nitrogen fixation and at the same time provide food, fuel and fodder, are particularly worthy of greater attention. Leucaena is already receiving worldwide attention. Several international conferences have recommended the collection and conservation of casuarina and neem germplasm.

Two conferences held on neem in West Germany in recent years have pointed out that the potential for improvement of the neem tree in yield and quality characteristics is considerable. India is participating in an international project to collect tree species used in arid and semi-arid areas. According to the list of priority species that has been prepared for this project, first prority will be given in India to three indigenous species (Acacia nilotica, Acacia senegal and Pros popis cineraria) and several exotic species of acacia, prosopis eucalyptus, and leucaena.

Among the second priority indigenous species are Albizza lebbek, pongamia pinnaka (karani) and Azadirachta indica (neem). Among exotics given second priority is Prosopis juliflora, which was introduced into India nearly a century ago but which is now widespread and shows considerable variation of form. Indigenous species listed for future consideration are hardwickia binata, Kingiodendron pinnatum and Dendrocalamus strictus (bamboo). Born completely blind, the Gangetic dolphin (okatabusta gabgetuca), remains sightless for the rest of its life. Nevertheless it does get by perfectly well, having adapted to a sightless existence in its murky, river habitat. Not to be confused with marine dolphins (Dephindae), the fresh water dolphins (Platanistidae) have lost all contact with the sea. They search for fish crustaceans and other such organism by sweeping the river bed with their sensitive muzzles.

Fresh water dolphins must come up for air every one or two minutes. During the day the dolphins perfer to stay in deep water, coming out at nightfall to feed in the shallows. Hydroelectric projects and the large fish nets of fisherfold pose considerable problems for fresh water dolphins but in recent years positive steps have been taken to identify and protect some of their known habitats in the Brahmaputra, Indus and Ganges river. Civets are cat-like creatures with elongated bodies that resemble those of weasels.

Solitary and nocturnal, these carnivorous creatures spend the day in hiding, coming to hunt small birds, mammals or reptiles at night. To accommodate their 32 to 40 teeth civets have evolved a longer snout which further differentiates them from the cat family. In India, Sri Lanka and even further east lives a civet that has come to be known as the 'today cat'. This creature, the Plam, civet (paradoxurus hermaphorditus) has a tail as long as its body and derives its name from its odd preference for fermented toddy. It often lives close to human habitation and, as many have discovered to their unpleasant cost, can give off an extremely foul, clinging odour form its anal glands if disturbed.

The palm civet is less carnivorus than its cousins and has specially adapted claws suitable for climbing the straight palm trunks. An ant lion larva digs a steep tunnel in loose earth and lies in waiting in the opit thus created with only its deadly pincers and head showing above the surface. A small insect which wanders too close, slides down the rim and cannot recover its balance as its legs get no purchase from the crumbly slopes.

The ant lion larva shoots up a spray of sand, which hastens the hapless victim towards its end. Having fed heartily on the digestible insides of its prey the tiny predators spits out the husk and patiently waits for the next victim. The Indian Pangolin (manis crassicadudata) has evolved a scaly body covering, fabricated out of modified hair. As protection against predators, it trolls itself into a ball and raises its sharp scales. These anteaters are mainly terrestrial, have no external ear openings and can seal their nostrils to protect themselves from ant bites.

Their long, extensile tongues are coated with a sticky secretion which allows them to capture ants, termites or larvae by probing deep into the labyrinathes of termite mounds which they rip apart with their strong claws. Pangolins have no teeth as their diet demands no chewing. A young pangolin will perch on its mother's back, just above the base of her tail, when it needs to be transported from one place to another. The great Indian hornbill (Buceros bicornis) is large, ungainly bird with strange nesting habits, the female seals herself into a hollow in a tree where upon her feathers soon moult.

The male dutifully brings her food for the entire period during which her eggs are incubated: feeding her through a narrow slit in the "prison" nest. When her young grow and their appetite increases, the female breakes down her prison doors to help the male obtain food.

9

Predators in Agroecosystem

Soil and water regimes have been altered as well in their structure and function as iii their dynamics. From an ecosystem point of view, agroecosystems resemble unmanaged systems by being solar powered, but differ because they have to rely on auxiliary energy to enhance productivity. Control of the system is external and goal-oriented.

Under natural conditions, a succession would start immediately after a disturbance, such as deforestation by a catastrophic event, which would soon become a species rich ruderal community and within centuries a forest. When man clear-fells an area for planting, he does not allow natural development, but restricts duration of succession and rigidly controls the pioneer vegetation by planting only one species, which is the crop. With each crop, a controlled short term secondary succession is started.

Early successional stages are nutrient-rich, with a high risk of nutrient loss when incorporation of nutrients into biomass by soil organisms is limited. Above and below ground species richness is low in the early stages of succession when a few specialized pioneer colonizers with high dispersal and reproductive capacities have the opportunity to build up large populations. Diversity soon increases and may, in the intermediate stage, reach even higher levels than in the final successional stage. These general observations also apply to the small soil animals, the soil mesofauna.

The importance of a rich soil mesofauna for ecosystem functions has been documented in many publications on faunal-microfloral interactions and nutrient cycling. Not only food web and nutrient cycling, but also the structural aspects of soil are positively influenced by these small animals. The feeding interactions of arthropods in soil are reviewed by Werner and Dindal. Apart from pests, the effect of soil microarthropods on plant productivity is mainly indirect.

Although the predators among them use only a very small amount of the system's energy, they may influence primary production considerably by controlling potential pests and by taking part in the rate regulation of nutrient cycles driven by the decomposer chain. Agroecosystems are frequently disturbed and succession is started anew. Consequently, for understanding agroecosystems, ecological studies from early successional stages are of great value. When a field is prepared for a new crop, ecosystem development is controlled to a large extent by abiotic external factors.

With the establishment of populations interactions within the system become more and more important. Nematodes, for example, have been found to be fast colonizers and their predators soon follow and interact with their prey populations. General predator-prey models support the observation that control by predation is very efficient, particularly at low prey levels, even for prey with very high reproductive potential. Mites from the group Gamasina are one important group of predators involved in these internal control mechanisms. The above-ground structure of the vegetation layer greatly influences soil mesofauna abundance and diversity. This has been proven in field and mesocosm experiments. The effects are indirect by providing a favorable microclimate for the development of soil fauna, by influencing their food, e.g. microflora and by increasing vertical and horizontal patchiness and habitat diversity.

The importance of diversity for decomposition has been shown by Verhoeff and Brussaard, who concluded that different assemblages of species may be involved in the same process. Lawton discussed several hypotheses on the implications of species richness and ecosystem function (necessity of minimal diversity beyond that redundancy; all species contribute to ecosystem performance; ecosystem function changes unpredictably when diversity and/or species composition changes).

The impact of agricultural practice on soil biota and their functions very much depends on the cultivation method chosen. Ploughing buries the organic material into deeper soil, where under well aerated conditions fast metabolizing bacteria may thrive., whereas no-tillage favors the development of fungi. Thus, the decomposer food chain in a conventional tillage system is bacteria-based, with a high risk of loss of nutrients to leaching. In a no-till system it is fungi-based and more nutrient conserving. For soil biota, ploughing is an elementary catastrophy and causes loss of diversity and population density. In contrast to conventional cultivation with its rigid external control mechanisms, integrated, sustainable or 'alterative' agriculture recognizes the advantages of natural processes and tries to protect soil life and its diversity.

Availability of organic material and a favorable microclimate may be provided by appropriate weed control or mulching. In many cultures, e.g. vineyards, management of habitat diversity in terms of a two or multilayered vegetation structure has been successful in increasing microarthropod diversity. Concepts of integration provide the tool for lowering energy subsidies and increasing ecological diversity. Tillage techniques are among the most important management tools for achieving sustainable productivity. An ecological agriculture is only possible in a sound economical framework.

The economic benefits of an integrated agriculture are well documented by Pimentel. The assessment of the importance of diversity and the presentation of solutions to the very specific problems of individual farms still requires tremendous research effort. Even mere documentation of diversity may cause difficulties, since it is based on counts of numbers and identification of systematic groups (taxa) of organisms. A specialist will be able to estimate diversity at species level, and will find high values compared with an other researcher who counts on the level of an order, such as mites or spiders.

Therefore, the determination of organisms (taxonomy) is a prerequisite for studying diversity. The value of lumping together species in functional groups or guilds is very much questioned by the work of Walter and Ikonen. With overwhelming species richness and high abundances within the range of hundreds of thousands of individuals per m, mites and Collembola are the most important microarthropod groups. Among the mites, predatory Gamasina are a group which has an important controlling function and which may be used for indication of soil quality and anthropogenic impact.

Mites (Acari)

Mites can be found in virtually all habitats, front the sea to the glacial zone, and in the soil, even in ground water. Almost any food source is used, and a great number of species are ecto or endoparasitic. This remarkable diversity is reflected in numerous morphological and biological specialisations: 40000 mite species have been described tip till now: many more are expected in the future. By their high diversity and often great numbers, they are deeply involved in many ecological interactions. In the last decades, they have attracted increasing interest in the context of decomposition, pest problems and their control, bioindication and human health.

With body length of rarely more than 2mm, the mites are the dwarfs among the spider-like arthropods. The small size not only has secluded them for centuries from close inspection, but also involves many specific morphological peculiarities. Like the other arthropods, they possess a chitinous exoskeleton, which may be soft or sclerotized to various degrees. Sclerotized parts of the limbs and plates of the body are connected by soft membrane. Hairs of various shape and number originate from the cuticle.

They are arranged according to specific patterns. Body division, number and shape of plates and hairs, and type of setal arrangement are important identification features. The order Acari is divided into the two suborders Acariformes and Parasitiformes. The Acariformes comprise three large systematic groups, the Acaridida (Astigmata), Actinedida (Prostigmata) and Oribatida (Cryptostigmata).

The -stigmata and -ida terminologies are only roughly identical. The Parasitiformes include three smaller systematic groups (Opilioacarida, Holothyrida,

Ixodida (ticks) and one larger group, the Mesostigmata or Gamasida. Apart from some external or internal parasites of animals, the Parasitiformes are for their majority predators. Mesostigmata possess one pair of stigmata connected with an anteriorly directed groove, the peritreme. Four free-living developmental stages are found: larva; protonymph, deutonymph; and adult. The group is divided into two systematic branches, one of which combine, the Sejina with the Uropodina and Gamasina. Sejina mainly live in tropical forest litter. Gamasina and Uropodina include many important species of agricultural soils all over the world.

Uropodina

Uropodina are found in large numbers in the range of 10000 m^{-2} in highly organic materials, such as compost, manure, older sewage sludge, decaying stumps or deciduous forest litter, where they prey mainly on Nematodes, insect larvae, unfit or even decaying earthworms. In less organic substrates, such as meadow or agricultural soil, Uropodina abundances are low.

Species number per site is generally in the range of ten or less. Distribution of Uropodina species is quite specific to habitat quality. Increase of pore volume and soil organic matter is reflected in rich uropodine communities. The preferred habitats of Uropodina (decaying organic matter with many Nematodes) are mostly ephemeral, and the mites have to migrate when their habitat has been decomposed. Only deutonymphs migrate. They exhibit a specific behavior before they adhere to insects which bring them to a 'fresh' habitat.

An adhesive stalk or tail is produced from the anus to attach to the carrier, a feature which is included in the scientific name of the group (uro: tail; pous: foot). Raut and Panigrahi suggest the use of Uropodina in biological control,: the mites have been observed in attacking slugs and killing them by making holes in the dorsal surface of their body.

Gamasina

The majority of the Gamasina are mobile predators. Gamasina have no eyes, but there is an overall sensitivity to light, which they avoid. They detect their prey by chemical or tactile stimuli. Volatile kairomones of spider mites, for example, are detected by Phytoseiidae. Most of the sensilla are located on the pedipalps and on the first pair of legs. The pedipalps probe the surface of the ground, apparently for orientation and chemical stimuli.

The first pair of legs serves as a feeler and is used to fix the prey by pressing it to the ground for piercing or cutting the skin with the chelicerae. Prey fragments are not swallowed, the predators digest their food preorally by producing digesting liquor, which dissolves the prey and is sucked up. Eggs may be produced sexually or parthenogenetically. A six legged larva hatches, which after moulting to protonymph and deutonymph develops to the adult stage with sexual organs.

Life history data vary considerably, e.g. development times from egg to adult have been reported from 2 (36°C) to almost 200 (25°C) days, longevity from 5 (20°C) to more than 300 days (15°C), for different species from various sources cited in Ruf. Information on temperature is important, since life history data are very much influenced by microclimate. In open landscapes, Gamasina occur in the range of 10000 m^{-2} in undisturbed meadows (0-15 cm depth) and even more can be found in disturbed soils or in early successional stages such as dunes or ruderal sites. Species numbers are in the range of 15 for undisturbed open grassland sites; ruderal sites may yield more than 25 species and forests more than 30. Gamasina are typically predators with the exception of some genera from the superfamily Phytoseioidea, which live in the herbal layer and feed on pollen. Epedaphic Gamasina are large (ca. 0.8-1.5 mm), long-legged species, well sclerotized and yellow-brown in color. They hunt on the soil surface in the litter layer for other small arthropods, particularly Collembola. To a large extent, they belong to the families Ascidae, Parasitidae, and Veigaiaidae.

Others feed on eggs of insects and their first instar larvae, preferably in habitats such as compost, animal droppings or manure (many Macrochelidae). Euedaphic species live in the soil pores. They are small and slender (ca. 0.25-0.5 mm), whitish in color and with rather short legs. The typical soil dwellers from the family Rhodacaridae have a divided dorsal shield, the anterior of which may be partitioned again to accomplish maximal flexibility in the narrow and crooked pores. Soil dwellers probably do not have much success in hunting in the pores by running. Rather, their hunting ability may depend on availability, i.e. on the density of prey in the microhabitat. Nematodes and other small wormlike food generally abound in the soil and are their preferred prey.

These animals are seized with the chelicerae when they stick their bodies into the air filled lumen of the pores. When there is mass occurrence of soil microarthropods, such as 'root mites' (Rhizoglyphus, Astigmata) or Onychiuridae (Collembola), these may also be eaten by Gamasina. Feeding preferences are widespread and even specialization occurs, as in *Alliphis siculus*, which exclusively feeds on Nematodes.

The form of chelicerae has been associated with feeding preferences by Karg. This view is supported by weak correlations of morphometrical parameters of chelicerae and food preference by Buryn and Brandl and opposed by Walter *et al.*

In a rural landscape with hedges in Southern Germany, Buryn and Hartmann found the highest diversity of soil Gamasina at the edge between the cultivated meadow and hedge, followed by hedge and meadow. Nematophagous species were dominant in the meadow and microarthropod feeders in the hedge. The hedge community included typical forest species. Also for other habitats or successional stages, specific species associations have been found, but autecological studies are too rare to find causalities for the existence of these communities. For recently disturbed sites with ruderal vegetation and low humus content *Arctoseius cetratus*, species of the family Rhodacaridae, particularly *Rhodacarus silesiacus* and some surface species, like *Parasitus eta* or *Veigaia nemorensis* are characteristic.

If such soils contain moderate amounts of organic material. *Alliphis siculus* is found regularly and Macrochelidae are typical colonizers of fecal droppings. Recolonization of disturbed sites very much occurs by wind drift and insect transport. As described for the Uropodina, phoretic distribution is also found particularly in deutonymphs of Gamasina species which colonize ephemeral habitats., such as dung; special morphological features of migrating stages do not exist.

Experiments showed that diaspores in topsoils even persist under treatment by heavy machinery, transport and desiccation. They can develop under favorable microclimatic conditions. Recolonization also may be fostered by hedges with their species rich edges and by islands with preserved sod in an otherwise heavily worked in agricultural landscape. Like other soil mesofauna, population development of Gamasina is very much influenced by microclimate, which depends on the structure of the herbacious plants, shrubs or trees and on the litter layer.

Removal of litter reduced species number in a forest from 21 to eight species. Experimental increase of soil moisture and reduction of daily temperature amplitude by cages covered with a 1 mm mesh raised abundances by a factor of five, which was mainly due to *Arctoseius cetratus*. This experiment was conducted on an early successional site and no major changes of species spectrum were observed. Gamasina do not change soil structure or plant productivity directly. However, as predators, they influence population growth of other organisms and through this have an indirect effect on overall ecosystem performance.

The multiple interactions with their prey are pretty diverse, not only with respect to their food, but also with respect to the soil profile and the patchiness of habitat. Nematodes play a central role as prey for Gamasina in agroecosystems. Nematode pests may be controlled very efficiently by these mite predators. Experiments by Bund document Nematode reductions of almost 70% due to predation by *Hypoaspis aculeifer*. Much evidence for the effectiveness of soil Gamasina in holding in cheek potential pests is reported by Karg.

Conventional agricultural systems are characterized by frequent and repeated disturbance, often within the mite's life cycle, and by 'clean' soil surface, particularly between harvest and cultivation. The depth of disturbance is considerable when a plough is used. As early as in the first half of this century. Franz published drastically lower abundances of soil fauna in fields, compared with uncultivated landscapes and on the consequences for soil fertility. The same trend has been observed by various authors for Gamasina. Compared with forests, Gamasina species diversity in conventional fanned fields, may be reduced by more than 75% to species numbers in the range of five.

Besides the mechanical disturbance, this drastic impoverishment can be related to the lack of a litter layer, which does not allow the development of a diverse prey spectrum to feed the predators. As in 'natural' sites, specific communities are associated with the peculiar environmental conditions of agricultural fields. Survival of ploughing or other structural 'catastrophies' (from the point of view of a soil organism) or fast recolonization with high tolerance towards microclimatic fluctuations in bare or sparsely grown fields, are some of the selecting requirements. Small euedaphic species with rather high reproductive capacity are able to cope with these conditions, e.g. Rhodacarellus silesiacus, Arctoseius cetratus.

These species are found not only in fields but also in early succession of ruderal sites, in sandy soils low in organic material. Fertilization with manure or dung allows colonization by phoretic surface species, e.g., fro the families Macrochelidae and Parasitidae. Lack of predation pressure by Gamasina may result in the development of pests of otherwise harmless organisms. In a 'clean' monoculture, there is nothing else to cat for these populations than the crop. Considerable damage to the farm's economy is inevitable, either as loss of yield or as cost for pest control, or both. In a 3-year study, El Titi found that ploughing has the most negative effect on soil Gamasina abundance and species diversity, compared with rotary and broadshare cultivator. Besides reduced mechanical disturbance of the soil, no-tillage cultivation does protect the organic layer of the field.

By this management technique, abundance and Gamasina diversity can be raised by roughly 50%. However, no-till cultivation has to rely on herbicides to control weeds. Many herbicides persist and may cause reductions of population levels, as was observed in relation to an old field site by Moore *et al.*, 1984. It must be noted that old field sites may be very species rich with high abundances.

The other effect is related to the potential pest, for which alternative food may be provided. It is not only tillage technique that has an effect on Gamasina diversity, but also soil compaction by machinery. Heisler performed experiments with increasing compaction. Strong compaction in the wheeltrack reduced abundance of Gamasina by 40% and species number by 30%. Epedaphic Pergamasus species, Alliphis siculus and euedaphic Rhodacarellus silesiacus are affected most. Arctoseius cetratus is resistant to compaction and becomes by far the most dominant species. Such a high numerical concentration to few or even one species reduces the efficiency of the Gamasina community as controlling predators.

Mineral fertilization, particularly in combination with manure and moderate use of sludge with low heavy metal content has a positive indirect effect on Gamasina abundance by stimulating population development of prey organisms; diversity is not significantly affected. Pesticide application triggers complex reactions of the soil zoocenosis. Gamasina as predators integrate effects, which may be indicated as changes of abundance and species spectrum. The mites may suffer from direct toxic effects or indirectly from changes in the prey populations. In an experiment, the succession of Gamasina in the soil of two synchronous experimental plots (undisturbed succession, sown grass with regular mowing) was observed with and without initial Aldicarb application. On the treated sites, species numbers did not recover from the impact within one year and community structure was strongly concentrated to only one or two species.Of the controls, dominances were not so pronounced. A variety of species specific reactions were observed for Gamasina: total extinction; late recoverage; effect depending on type of vegetation; stimulated develop merit; and indifference.

Similar to the results of EI Titi, Rhodacaridae were most affected, which is related to the water solubility of Aldicarb. Stimulation of surface dwelling species may be due to an altered competitive situation and to increased hunting success because of slow, only partially poisoned prey. The experiment also confirmed indirect effects of reduced predation pressure on prey: since the abundance of Rhodacaridae was strongly reduced after the pesticide application, a collembolan prey population *(Mesaphorura krausbaueri)* developed unchecked. Another indirect effect may be possible causing the strong decrease of the nematophagous *Alliphis siculus:* Aldicarb is very nematicidal and *Alliphis siculus* suffers from lack of food. From the feeding biology and from their interrelations in the ecosystem, main areas of pest control with Gamasina can be identified: Nematodes; Collembola; insects; and pests of the shrub layer.

Since the early 1960s, Gamasina have been seen as an efficient biological control of Nematoda. Above ground, Gamasina of the family Phytoseiidae are important antagonists, particularly of spider mites. In this field much experience has been acquired, e.g, on the right timing of the release of the predator. Macrochelidae feed on eggs and larvae of insects. In poultry houses, housefly populations have to be controlled. Macrochelidae have been shown to effectively control these insects, when some prerequisites are observed, such as alternate prey to overcome lack of main prey, and leftovers of manure for Macrochelidae to breed. The few examples show that successful biological control must be based on good knowledge of the biology and ecology of the species involved. In a monograph on Gamasina, Karg gives an introduction to the group and a key to more than 1000 species from Central Europe. Either intact or mixed soil samples are collected from the field.

Care should be taken to separate minor habitats, such as soil horizons and patches of organic matter, distance to crop plant, within and between planting rows, etc. Gamasina are expelled from the soil by dynamic extraction, which takes advantage of the mobility of the animals. The principle of the extraction method is simple. In the laboratory, the soil conditions should be made comfortable for the animals, so that they are active (humid and cool). For mixed samples the soil should be handled gently to avoid crushing the mites.

The soil is placed into a sieve over a funnel and dried carefully over a couple of days by applying increasing heat, e.g. with a dimmable electric bulb. By this method, a moisture and heat gradient of ca. 10 K should be achieved between the surface and the bottom of the sample. Towards the end of extraction, this gradient is reduced and final temperatures of ca. 400°C at the bottom of the soil sample should be reached to make mites leave this zone.

The soil animals fall via the funnel into a glass with preservation fluid (70% ethanol) and can be counted with a stereoscope. Cleared specimens are needed for identification with a microscope, and permanent mounts are very useful. The distribution of soil animals is aggregated and high variation of abundance estimates in the range of 60% of the mean are to be expected (P= 90%). These values have been derived torn the analyses of extended data sets for a widely used sample size of n = *10*. Assessment of species spectrum is more reliable. Gamasina are caught alive either with dynamic extraction or in the field with a fine brush.

10

Threat of Biodiversity Loss

Data from the last three decades demonstrate that on a worldwide scale the total area under forest and woodland has remained fairly stable. Considering the conversion of natural forests in many parts of the world, this means that in general nature natural forests are being replaced by woodlands which are the result of human activity, be it planting of trees or the development of secondary forests on land which has been cultivated in the past. With the increasing demand for land, and for wood worldwide, this trend is likely to continue for decades to come.

Consequently, it can be expected that forests which bear the marks of human influence will play an increasingly important role in species conservation, particularly in tropical countries where the largest biodiversity must be preserved, but also where economic needs put considerable pressure on natural resources. This is being increasingly recognized by forestry officials and forest managers worldwide. Many countries in the temperate zones which no longer have any significant amount of mature natural forest manage biodiversity in anthropogenic forests.

In many European countries there is only little need to generate economic returns from those forests, while in the USA the National Forest Service recognizes a serious need to balance economic demands with biodiversity. In many tropical countries where forests are managed for economic interests, maintaining a diverse forest with a large number of species is often considered contradictory to the principal goal of increasing returns. This, for instance, remains the case with most of the plantation and land development programs that the Indonesian Ministry of Forestry has designed for areas of forest land which nave been designated Conversion Forest. Indonesia has a total land area of 199 $\times$ 10^6 ha, including about 10^9 $\times$ 10^6 ha of forest and woodland of which 86.4 $\times$ 10^6 ha are forest cover. This is about 6% of the world's total remaining tropical forest, and the third largest area after.

Brazil and Zaire, Indonesia's forests contain about 10% of the world's plant species, 12% of the world's mammals, 16% of the world's reptiles, and 17% of the world's bird species. Because of its unique geographic location on both sides of the Wallace Line, between the continents of Asia and Australia, a high rate of species endemism occurs. On the other hand, Indonesia's forests constitute an important economic resource for the country. Until recently, timber was the second largest generator of foreign exchange after oil, and remains an important source of development capital Not with standing this economic importance, Indonesia has repeatedly expressed an awareness of it's responsibility to the world community to preserve it's forests and the biodiversity they contain.

In order to safeguard both the important economic role of the timber industry and the country's biodiversity, Indonesia has functionally divided its large forest. This division was laid down, until 1996, in the Forest Land Use by Consensus which is now being replaced by the Provincial Land

Use Ordination Plan. Roughly one-third of the country's forest lands are set aside for conservation or protection purposes. The larger part is designated as Production Forest, while the remaining area is designated as Conversion Forest or Nonclassified Forests. Conversion and Non-classified Forests may be used for agricultural or other purposes.

The total of the previous forest land categories is 144×10^6 ha, a result of the fact that not all forest land actually has forest. The Government of Indonesia considers one of the major threats to the timber-rich forests of Indonesia to be the many swidden agriculture farmers in the outer islands where the larger areas of Indonesia's forests remain.

Hence, the GOI has made it a goal to eradicate all forms of swidden agriculture in the whole country. It has designed several programs to reach this goal. These programs vary according to the category of forest in which they are being conducted. So called 'social forestry' programs have been designed to stop forest encroachment by small farmers in Conservation, Protection, or Production Forest areas.

In forest land that is designated as Conversion Forest, swidden agriculturists are encouraged to participate in estate crop development programs, especially rubber and oil palm, or to become wage laborers in commercial timber estates. In this chapter we will discuss the programs that the GOI has been developing for areas of Conversion Forest, and juxtapose them to a case of traditional forest management by swidden farmers in West Kalimantan, one of the four Indonesian provinces in Borneo. The four provinces of Kalimantan hold 44×10^6 ha of forest land, or 31% of the total forest area of Indonesia. In 1990, 34×10^6 ha still had mature standing forest. 11×10^6 ha are permanent Conservation and Protection Forest, 26 X 106 ha

Production Forest, and 8 X 10^6 ha Conversion Forest. A total of 4.4 X 10^6 ha are estimated to be under swidden agriculture, which sustain 228 000 households, or 17.2% of the total population. The forest cover in Kalimantan is estimated to disappear at a rate of 0.49% year^{-1}.With the present discussion, we hope to demonstrate that the programs designed to reduce the threat that swidden agriculture supposedly poses for In donesia's forests will necessarily lead to a significant reduction of the high biodiversity typical of many areas which are, predominantly, under swidden agriculture.

Furthermore, we argue that alternative choices for economic development of swidden agriculturists can be designed which can preserve, or even enhance the biodiversity in these same areas. In particular we will try to demonstrate: (1) that conservation of biodiversity is an integral part of resource management among the very same swidden agriculturists as an important subsistence strategy; (2) that several of the programs which are designed to eradicate swidden agriculture in Indonesia ironically will lead necessarily to a loss of this biodiversity that is maintained by swidden agriculturists; and (3) that instead of promoting large-scale timber plantation or estate plantations, developing traditional forest management is both an attractive option for regional economic development as well as a way of preserving the biodiversity that is an integral part of many swidden agricultural landscapes.

It is estimated that swidden agriculturists are responsible for about half of the estimated 1 X 106 ha land deforested in Indonesia yearly, although little accurate data exist to sustain such an estimate. Of the total area under swidden agriculture, about 25% is in Production Forests, 15% in Protection Forests, 29% in Conversion Forests, and 31% in Unclassified Forest. Even though 60% of swidden agriculture is located in forest land which is planned to be used for non-forest purposes, swidden agriculture is considered to have serious implications for the future of forestry in Indonesia.

Since the early 1970s, the GOI has designed programs to deal with swidden agriculturists. In their earlier phase these programs attempted to relocate swidden agriculturists to out of forest areas. Especially since the mid-1980s, when the issue of sustainable forest utilization has become widely discussed in Indonesia, many directives from the Ministries of Forestry, Agriculture, Transmigration, Social Affairs and Home Affairs have appeared which are meant to 'stabilize' swidden agriculture.

The social forestry, or village development programs, which are applied in Conservation, Protection, or Production Forest areas, aim to give people responsibility for managing the forests in their areas. In fact social forestry programs are still ill-defined, and subject to much discussion. Experiences with social forestry in Java, where local control over forests was replaced by government control some time ago, give little hope for similar programs for the outer islands.

Social forestry is not attempted in areas which have been designated as Conversion Forest. In this category of state forest land three different programs have been defined: (1) the establishment of industrial timber estates ; (2) the nuclear estate program and (3) the Rehabilitation and Expansion of Export Crops program. The timber estates development program began in 1984. This program has a high, priority with the GOI, since in the near future, it is supposed to substitute for the expected decline in supply of timber as a result of exhaustion of the natural forests. Until 1988, some 1.4×10^6 ha of timber production plantation existed. The largest parts of those timber estates are still found in Java, but the second largest parts are located in the four provinces of Kalimantan with most in West Kalimantan Indonesia's goal is to reach 6 × 10^6 ha of industrial timber plantation, by the year 2000. These plantations are to be established in Conversion Forest or Production Forest areas if they do not fall under a forest concession and do not contain sufficient commercial natural forest.

In practice many state HTI projects have been established in secondary forest, scrub land or grass land. The location of HTI estates in secondary forest areas implicitly means that it is located in land that is occupied by swidden agriculturists. One of the objectives of HTI estates is also to rehabilitate degraded lands. However, often any land that is under fallow is qualified as degraded by forestry officials, although from the swidden agriculturists point of view this land still may be productive rice land.

HTI planners give little consideration to the actual importance of this rice land for swidden agriculturists. The HTI estates are largely to be planted with a few fast growing species. For instance, INHUTANI III the state company in charge of establishing the industrial forest plantations in West Kalimantan, has been planting only *Pinus merkusii* in the HTI estate in the district of Sintang, in which eventually 100 000 ha of land are to be planted. In the neighboring district of Sanggau about three times !bat area are to be planted to *Acacia mangium.*

The participation of swidden agriculturists in industrial timber estates is mainly as providers of labor and land. Land needed for the estates is obtained from local farmers who are given some compensation, but this is not higher than US$15.00 ha. In the original wans the timber estates were to provide work to the local population, hence generate income, and thus reduce the need for swidden farmers to grow their own food, and slash more forests.

The wages paid are rupiahs 3750 day 1, or US$1.70. Only when the implementation of industrial timber estates is combined with the Transmigration Program are housing infrastructure, schools, medical posts, and the like, provided to the transmigrants coming from Java, Madura, Bali, and Lombok, as well as local farmers who agree to participate. HTI/TRANS projects aim to include about 50% of 'local' transmigrants in their schemes. The NES/PIR or Nucleus Agricultural Estate Program, which began in 1979, and which is the responsibility of the Ministry of Agriculture, aims at establishing estate crops. In this program government or private companies establish an estate, which may be as large as 50 000ha. About 60% to 80% of the land is dedicated to smallholder production of the estate crop.

Participant farmers are given 2 ha of land to establish the estate crop, and 0.5 1 ha for a house lot and a home garden. The estate company provides facilities to the participating farmers to establish their 2 ha of estate crop, and it guarantees to buy the yield, as well as to provide technical assistance and credits for inputs.. Farmers accept these facilities as a loan which they have to pay back. In West Kalimantan farmers must repay US$4900 plus an interest of 6%. Farmers who become estate members commit themselves to work as laborers on the company-owned part of the estate, for which they receive daily wages.

Incomes that participating farmers receive from the sale of oil palm fruits are supposed to climb from RP37 500 month after 3 years to RP 220 000month1 after 12 years. These predictions, however, do not agree with recent actual earnings after 6 years of only RP88 000month1 that farmers in some Nuclear Estate Programs obtained. A new version of the same program is the PIR/TRANS program. This also establishes nuclear estate crops, but largely with the participation of transmigrants coming mainly from Java. Bali and Lombok.

In many PIR/TRANS projects local swidden agriculturists are incorporated also, but since the estates rely on transmigrants for their labor, fewer local farmers can be involved which means that relatively more farmers will be resettled. Farmers usually do receive compensation for the land they give up. Reports, however, document the coercive, methods used by the state companies to obtain traditionally held land. NES/PIR or PIR/TRANS estates are entirely converted to mono-crop plantations, except for a smaller part which is converted to food crop area.

The plantations used to be slash-weeded frequently, but increasingly chemical defoliants such as Monsanto, Roundup and Kleenup are used, and under the mono-crop only a grassy weed vegetation remains. Hence, the claim that NES/PIR or PIR/TRANS projects aim to conserve the environment contrasts sharply with their actual outcomes. In West Kalimantan, the larger part of NES/PIR or PIR/TRANS plantations are under African oil Palm and smaller areas under rubber plantations.

In 1993 the total area under oil palm plantation in West kalimantan was 142 403 ha the total area planned to be converted into nuclear estate crop is 1.2 × 106 ha, or 8% of the total surface of West Kalimantan. The third program that affects swidden agriculturists in West Kalimantan is the Rehabilitation and Expansion of Export Crops program. This program provides credit to farmers to improve the quality of their smallholder plantations. In the outer islands, support is given for 2 ha of export crop land. In West Kalimantan this program only focuses on improving the quality of rubber production. Very soon after, the first *Hevea* rubber appeared in Sarawak early this century, it also was brought to West Kalimantan by the Dutch.

Most rubber, however, has been grown as 'forest rubber', planted on swidden fields, and left to grow, together with the spontaneously, occurring secondary vegetation. As a result, most rubber gardens are diverse, species rich, anthropogenic forests. Farmers who receive credit and technical assistance under the Expansion of Export Crops Program commit themselves to plant only improved rubber clones. These are high yielding rubber clones, which are much less resistant to pests than the locally used varieties.

In many cases, how ever, farmers have actually been sold the locally existing rubber as improved high yielding rubber. Income levels again are much lower than predicted when farmers where invited to participate in the project. The numbers of trees on single improved rubber holdings varies between 160 and 800 trees with the highest number of 800 trees, after 9 years an income of Rp 120000 month can be achieved. This is significantly above the average rural cash income in the region, but must cover purchase of most household provisions. Farmers who plant improved rubber clones are supposed to conduct a thorough slash-weeding from the onset of the plantation until the trees are large enough to be tapped. As a consequence, improved rubber gardens have a characteristic mono-crop species composition which is similar to that of Pinus *merkusii, Acacia mangium,* or oil palm plantations and not like the diverse forest rubber gardens found in Kalimantan and Sumatra.

A dominant, mono-culture plantation has a grassy weed vegetation covering the forest floor. Since currently more than 400000ha of land are under traditional species-rich smallholder rubber gardens in West Kalimantan, the extension of loss of biodiversity can expected to be considerable if these areas are to be converted to monoculture improved rubber gardens.

The island of Borneo, with an area of 642 000 km^2 consists of the two East Malaysian states Sarawak and Sabah, and the four Indonesian provinces of East, South, West, and Central Kalimantan.

In 1990 the whole island had 12.7×10^6 inhabitants, 9.1×10^6 of which lived in the Kalimantan provinces. About one-third of the people who live on the island are broadly classified as indigenous Dayak. Throughout the history of Borneo the island has been of interest for its natural resources. In pre-Dutch colonial periods for various forest resources and minerals, and for the last several decades for its oil and timber.

On the other hand, agricultural products such as pepper and rubber have been important export crops since about the turn of the century. Production of oil palm also has been increased drastically in the last decades. The changing economic importance of the island also has resulted in considerable social changes. In the Indonesian part, until two centuries ago, it was mainly Malay who inhabited the coasts and river banks while indigenous inhabitants lived in the interiors.

Gradually various other groups came to claim their niche: the Chinese, the Bugies from Sulawesi, and later people from other islands from the archipelago, as transmigrants, or as spontaneous migrants, continuously pushing back the native Dayak to more remote places. Together with the booming of timber and oil exploitation came the large scale opening up of large parts of West, South and, to a lesser extent East Kalimantan. This opening up was followed by large scale agricultural development projects. The native Dayak population mostly living in the remoter interiors relies for its daily staple on upland rice cultivation, produced in yearly cleared swiddens. Much controversy and confusion still exists about swidden agriculture.

The term covers agricultural practices which begin with the slashing and burning of vegetation of an area of land, using it for intensive cultivation for one or several years, followed by fallowing the same plot for a longer period. Among most swidden agriculturists, fallow fields continue to have an economic importance since the secondary vegetation contains spontaneously occurring species, or perennial species which, were planted at an earlier stage and which continue to yield useful products.

Some of these species occur mainly in fallow vegetation, and are only exceptionally found in primary forest Jessup claims that some Kenyah farmers in East Kalimantan purposely maintain a sufficient area of fallow to assure the supply of such species. Hence, as a rule, among swidden agriculturists some kind of management of the fallow vegetation occurs. Among many swidden agriculturists the dominant economic imperative is to optimize returns of labor, rather than returns of land.

In order to achieve optimal returns per laborer, swidden agriculturists prefer to fallow their land, since this will avoid low yields front plots, or excessive need for labor for weeding. Therefore, although fallowing of agricultural land is not limited to the tropical regions, swidden agriculture is an adequate adaptation, especially to these hemispheres with naturally low soil fertility, and high rates of weed invasion due to a hot and moist climate, as long as land pressure remains low.

Swidden agriculture is still practised by many farmers in the world, and subsequently appears in,many different forms. Even on the island of Borneo, significant variations of swidden agricultural patterns can be observed. For instance, lban farmers historically had the reputation of preferring land that was newly opened in mature natural forests, and shied away from making swiddens in secondary forest, although Jessup and Sellato question the validity of such generalizations based on ethnic affiliation. Other groups, like the *Kenyah and Kenyah*, cultivated the area around a new settlement for several decades, and then moved the whole village to a new site.

These farmers cultivated land only a few times, as long as they stayed in one place, leaving it to recover for long periods after they moved. Often new settlements were constructed within less than one day's walking distance from the previous sites such that old fallows or forest gardens could easily be harvested when fruiting occurred.

A reason to move the village could be the exhaustion of agricultural land, or because diseases occurred at increasing frequencies, or in some cases the location had become unsafe as a result of invasions by other groups. Recent migrations have occurred because people wanted to gain easier access to markets and government services. The *Bidayuh*, a name originally used to distinguish the non-Malay, non-Iban inhabitants of Sarawak's first division, constitute several different linguistic subgroups, who originated from the Indonesian part of West Kalimantan, and Which are found as far South as the Ketapang basin.

Bidayuh farmers practice a type of swidden agriculture which is more sedentary than that of several other groups in Borneo who migrate more often. Rather than moving entire villages, Bidayuh villages which become too large, are split and only portions moved to a new area of forest to start their own settlement. The stable Bidayuh villages, however, frequently move the location of their settlement within the village territory, over distances that do not exceed about 1 km.

The more permanent character of Bidayuh settlements has certain consequences for the way swidden agriculture is practised. In order to be able to reside over the long term in a single region, a farmer must have a sufficiently large holding of land under various stages of fallow. This holding is built up gradually from when a couple become an independent farming household, until the area is large enough such that fields can be used and fallowed at sufficiently great intervals. Since *Bidavuh* farmers have more stable holdings, the area to which they hold tenure claims are smaller than, for instance, the ever-moving *Than.*

On the other hand, land tenure claims are strong, and are held by descendants for many generations, even after a settlements splits and one part of the community moves somewhere else. Another important feature which is typical of *Bidayuh* sedentary swidden agriculture is the emphasis on tree planting and conservation of forest resources. Although *Bidayuh* swidden agriculturists engage in frequent conversion of species-rich mature natural forest into species-poor agricultural land in terms of higher plants and larger animals, their resource management, in general, is very conservation minded. Honorable as this may sound, this is mostly a result of economic interest.

The landscape of a typical *Bidavuh* village is usually characterized not only by land under swidden and fallow, but also by many patches of forests of various sizes. For instance, among two different *Bidavuh* groups in the North of West Kalimantan, *Mate* Dayak and *Jangkang* Dayak, six types of managed forests have been described. These include communally held forest, or *hutan tutupan*, which are usually located close to villages and may be as large as 100ha. *Hutan tutupan* mostly function as a reserve for timber for house construction, rattans, medicinal plants, fishing and hunting sites, and to protect water sources. A smaller version of the same type of forest, of between 1 and 3 ha, is forest islands *or pulau rimba.* Such patches of forest become surrounded by one household's agricultural fields or fallows and according to local customary law, became private property of this household. These forests are often kept as private forest reserves, and yield the same products as *hutan tutupan*. In the same category of protected areas of mature natural forest fall the *sompuat*, which are patches of forest which surround honey trees, mostly *Koompassia excelsa* or *Koompassia malaccensis*.

The *sompuat* forests are kept intact in order to provide an appropriate habitat for the *Apis dorsata* honey bees that built their nests in the central tree. Obviously, preserving areas of mature natural forests also implies preserving the diversity of species that are present in them. However, that is not where *Bidayuh* biodiversity forest conservation stops. Two types of managed forests are important, riot only economically, but also for the diversity of plant species they contain. These are *tembawang* or fruit forest gardens, and rubber gardens created by individual farmers planting trees on sites previously used for agriculture or around the settlement area, or sometimes directly in mature forest.

Planting may continue for many years until eventually most of the area designated for tree planting is covered. Such *tembawang* are privately owned and inherited by descendants of the people who established the forest. They are usually about 1-2ha in size. Since many farmers in a single village create individual *tembawang* on adjacent fields, over time these merge into a single stretch of contiguous forest. Because of the short distances over which the village is frequently moved, the *tembawang* area often trails the advancing settlement as a long band of high closed forest. Rubber gardens are usually located adjacent to *tembawang*. Once the rubber seedlings are planted, no weeding occurs until the rubber is fully grown; about 10 years.

Only then are trails established to reach trees which have to be tapped. Very often additional fruit trees are planted in rubber gardens, as a result of which they tend to become an extension of the *tembawang* area, rather than a totally different type of managed forest. The biodiversity conservation role of managed natural forests like *hutan tutupan, pulau rimba,* and **so**mpuat are obvious. However, in this respect tembawang and rubber gardens are no less important.

Although the principal products planted in a tembawang are fruits, bamboo, rattans, ironwood illipe nut or other dipterocarp species, many other naturally occurring species appear spontaneously in tembawang forest. Often, if they are of economic interest, these species are protected or tended, while other species are tolerated and harvested as needed. Mature tembawang are full grown forests, which have a structure that is similar to that of mature natural forest and which hold a considerable biodiversity.

In one of the villages researched, for instance, 400 plant species were recorded in five *tembawang.* These five *tembawang* had an average of 170 plant species per hectare, while one specific *tembawang* had a total of 231 plant species per hectare. In two plots of mature natural forests in the research area 256 and 258 plant species per hectare were recorded. High species diversity can also be found in rubber gardens.

The proportion of such managed forests in the landscape further stresses their importance for the conservation of biodiversity. In one village we found that of the approximately 1800ha under cultivation 600ha are tinder species-rich tembawang or rubber gardens. It should be evident that any of the three programs that the GOI wants to implement in Conversion Forest areas will necessarily result in a considerable reduction of diversity in the landscape and the biodiversity it contains. From the biological conservation standpoint therefore, it would be better to leave swidden agricultural land the way it is, rather than try to implement large-scale plantations of mostly single species. However, the objectives of HTI, NES/PIR and NES/TRANS and export crop development are not primarily to conserve biodiversity, but rather to increase or sustain the state's income from timber production, or estate crops.

These incomes are indeed intended to increase living standards of local fanners, but also enhance the country's timber, oil palm and rubber industry, in order to provide income for the state and to generate foreign exchange. The programs that deal with swidden agriculturists in Production, Protection, or Conservation forests, focus largely on developing alternatives which reduce the reliance of farmers on forests as a resource to create agricultural land. Some of the strategies are to set aside forest land from which products can be obtained for sale in local markets to obtain monetary income.

The programs reserved for areas of Conversion or Unclassified forest land do not consider options of forest exploitation, although they are required to dedicate some land to other uses. For instance, the HTI estates have to dedicate 10% of the land to *tanaman kehiduppi*, which can be private vegetable or mixed fruit gardens. Similarly, PIR projects provide land to participating farmers for their own house gardens. Considering the objectives of estate development, alternative choices will have to provide the same incomes as the estate options to be socially acceptable. They have to he generated at least partly from some kind of exploitation of forests in order to be able to choose alternative approaches for developing swidden agriculture without losing its intrinsic biodiversity.

Diverse tree gardens can provide significant *levels* of income, demonstrated by the way rubber is still produced in species-rich gardens, typical for most smallholder rubber producers. In many isolated villages in Kalimantan, rubber remains the most important cash crop and low harvest levels may be as much the result of low yields, insufficient trees, or simply a limited need for cash. This is riot to say that rubber production should not be improved. However, rather than trying to replace all traditional smallholder rubber gardens with improved rubber, or with nuclear estate rubber, replacement with clones that provide higher yields but which can be grown in traditional diverse rubber gardens, or in complex, tembawang-like forest gardens, should be considered also. Besides improving rubber yields from diverse species-rich rubber gardens, many other options are available to develop swidden agriculture without having to opt for monoculture production.

Reports exist from *Bidavuh* farmers in central West Kalimantan who gain substantial income from *durian* yields from *tembawang* forests. These fruits are being bought by traders from Kutching, Sarawak who travel for 6 h to reach the producers during the fruiting season. Expanding markets in urban Indonesia, Malaysia and Singapore only will increase the demand for such products.

Hence, the opportunity to increase incomes by producing durians in diverse forest gardens only will increase. Existing forest management practices, furthermore, provide a unique opportunity for smallholder timber production which eventually could sustain the supply of the local timber industry. Such timber can be planted in and harvested from rubber gardens, *tembawang, pulau rimba* and *hutan tutupan.* The 33% of area under forest in the village where Dayak forest management was studied suggests the potential that such traditional forest management holds for the production of high quality timber.

Some kind of tree cover creates a better environment to plant species which are naturally found in closed forests, rather than in open terrain where many seedlings die, or where fires may destroy plantations. Additionally, in diverse holdings, other faster-growing species could be produced on small lots which would be rotated, for instance, with rice production in order to supply an industry that needs faster-growing species of lower wood quality.

11

Establishing Research

According to Pandolfi, a correlation exists between biogeographic pattern and geological history, and modern marine distribution patterns can best be interpreted by incorporating the geological history of the area under study. Numerous hypotheses have been proposed to explain patterns of biodiversity on coral reefs. These include dispersal models, the pacific plate vicariance theory, an expanding Earth theory and a variety of ecological explanations.

However, the systematic study of comparative levels of endemism in coral reef inverte-brates to see which paradigm, or combination of para-digms, best explains biogeographical patterns remains to be engaged by the scientific community. Fundamental knowledge of biodiversity at the level of species is a prerequisite for such investigations. Integrated studies of marine biodiversity are just beginning. To date, many taxonomic and most biogeo-graphic studies of reef systems have been random and opportunistic, focusing on easily accessible islands.

Taxonomic literature on tropical marine invertebrates is scattered, dealing mostly with large, spatially obvious components such as fish, scleractinian corals, and molluscs, thus creating a biological bias for larger organisms that may not be the best indicators or measures of biodiversity. Platnick refers to the condition of concentrating on the obvious spatial components as the "megafauna bias." The most informative invertebrate groups in terms of biogeography are those without a dispersive larval stage.

Wide dispersal capabilities can mask small-scale distribution patterns. Thus, animals with restricted distributions more accurately reflect levels of endemism. Information on those groups of invertebrates with restricted distributions is needed to integrate the many competing hypotheses on distribution mechanism. While coral reefs contain the highest levels of biodiversity in any marine ecosystem, virtually nothing is known about possible extinction or natural trends in biodiversity.

World-wide reports of changes in coral reef systems are largely anecdotal and based primarily on ecological research. Human impact is suspected as a contributing factor in these changes, but the interaction of natural change versus human-induced impacts remains speculative. The rate at which marine areas now are being designated for protected status as sanctuaries, refuges, preserves, parks, etc., creates a problem in setting priorities an selecting among the many candidates suggested for protection. The goal of maintaining native levels of biological diversity requires that knowledge regarding historical sources and levels of biodiversity be known so that competent decisions about research, conservation, and management action can be made.

Distinguishing types of biodiversity

The marine environment is more diverse than

terrestrial systems, yet most of what we know about biodiversity comes from studies in tropical moist forest systems. Biodiversity studies in marine environments, particularly the tropics, suffer from diffusion of scientific effort and an inability to identify critical sites based on levels of biodiversity. Present management and conservation efforts in marine systems are driven mainly by concepts other than biodiversity, e.g., species of special human interest, areas of spectacular natural beauty, and economics.

Such an approach places the highest priority on unusual areas with little regard to levels and trends of biodiversity. In some instances where biodiversity surveys have been conducted, emphasis is on "spatially obvious" organisms such as fish and corals that may not be the best indicators to identity the processes of environmental change. Successful management of marine ecosystems must rely on relevant information about levels and trends of biodiversity, information that is almost entirely lacking in current management approaches.

Biodiversity survey and inventory programs that are designed to identify and select those marine areas of highest scientific value are needed. Resources such as existing collections in natural history resumes and coordination of ongoing taxonomic surveys and inventories need to be organized according to selective criteria. The limited number of active field systematists and taxonomists currently working in marine environments must be deployed effectively.

Training programs to allow a wider dissemination of taxonomic expertise also are a critical element in a unified approach to marine biodiversity. Use of data from disciplines outside taxonomy and systematics must be incorporated into a marine biodiversity program for the twenty-first century. Information from fields. Such as plate tectonics, paleogeography, cladistics, ecology, and anthropology must be brought together in concert to fully understand what is happening in marine environments and what can be done to improve our understanding of system processes. At the administrative level, specific shifts in policy goals to include reliance on preventative rather than reactive management strategies must become a priority.

Taxonomy and biodiversity in coral reefs

Available taxonomic data tell us more about varying attention given to different groups of animals, the "taxonomy of taxonomists," than about the level of taxonomic knowledge. The information summarized by May illustrates the great disparity of attention received by different groups, Roughly one-third of taxonomists work on plants, while the remaining two-thirds split roughly equally between invertebrates and vertebrates. The estimated total number of species of vertebrates is 40,000; species of plants is 300,000; and species of invertebrates is about 1 million.

Therefore, for every n taxonomists working on vertebrates, there are $0.1n$ taxonomists investigating plants and $0.01n$ taxonomists specializing in invertebrates. When we consider that a majority of invertebrate taxonomists study a single group, the insects, the great disparity within the current taxonomic work force that specialize in marine invertebrates becomes apparent.

Documenting biodiversity in tropical marine ecosystems

Despite extensive reports of large-scale change in coral reef systems, the scientific, conservation, and manage-ment community does not currently have the capability to investigate every region or problem that gains public attention. Lack of comprehensive marine biodiversity protocols to identify areas of particular scientific, conservation, and management value have lead to a diffuse approach to the systematic investigation of coral reef biodiversity.

Most research taking place now on reefs is ecological in nature, with little or no systematic effort directed toward taxonomic surveys arid inventories of reef systems worldwide. This lack of focus is further compounded by the shortage of experienced field systematists and taxonomists. Training programs that would increase the number of personnel available for surveys and inventories on coral reefs, while frequently discussed, have resulted in few active programs.

The need is especially acute for the marine invertebrates, particular groups of which arc sensitive indicators of change in coral reefs. The scientific community must develop a process to assess biodiversity that uses selective criteria and sets priorities, in effect a form of environmental triage. Every reef system cannot be investigated with current personnel and levels of funding. Difficult choices must be made that maximize existing personnel equipment, and organizational structure.

Numerous coral reef sites already enjoy protected status, while numerous additional sites are being suggested for protection. In many cases, the site selection process is driven from a perception that 3 particular site is in a stressed or deteriorating condition. This approach focuses limited resources on a series of crisis situations arid perpetuates a reactive management posture. Ideally, the scientific community should provide theoretical guidance and taxonomic information regarding appropriate site selection. 'The conservation community, capable of rapid response and adept at raising public awareness, then should use scientific data to target specific sites or regions for protection.

The resource management community does not become established until an administrative structure is created. Every effort should be made to incorporate high-quality science in management approaches that document biodiversity. Coral reef sites that enjoy preexisting protected status should be encouraged to adopt programs that adequately document levels of biodiversity if such information is lacking. These major groups scientists, conservationists, and resource managers must work in concert along established guidelines to maximize the application of diffuse resources arid funding to help identify and protect coral reefs.

Establishing priorities in marine biodiversity

Priorities vary depending on individuals, agencies, arid processes involved. For example, taxonomists place the highest priority for research on reefs that contain the highest biodiversity at the level of species or are the least impaired or impacted by man. Conservation groups may target areas that house species special interest or support high proportions of endemics, while resource mangers might focus on reef areas that seem threatened by human impacts. Any system of establishing priorities must take into account a variety of factors.

Developing selection strategies

The following approaches should be employed to develop priorities for marine biodiversity initiatives:

(1) Identify centers of evolutionary diversification or major centers of genetic biodiversity by surveys

and inventories, existing museum collections, and cladistics.

(2) Target areas where geological history indicates a history of vicariant events that might lead to high levels of endemic species or areas of composite biodiversity.

(3) Implement surveys and inventories to determine biodiversity baselines. While entire protected areas must be surveyed initially, ongoing monitoring of biodiversity could be restricted to special protected areas, such as replenishment zones located within a protected area, it is important to stress that our understanding of biodiversity in coral reefs is rudimentary; therefore, initial surveys must incorporate a representative cross section of the entire area. Once levels of biodiversity have been determined, decisions can be made as to the location of permanent monitoring sites and schedules.

(4) Publish comprehensive identification guides to biota. Most information regarding marine invertebrates from coral reefs is uneven in terms of taxonomic and geographic coverage. Computer-based identification manuals should be a priority for development and distribution.

(5) Establish parataxonomic training programs to argument survey and inventory capabilities. Basic biodiversity surveys and inventories of many protected areas remain to be initiated and are not currently a primary management concern. Two of the largest marine protected areas in the world, the Florida Keys National Marine Sanctuary and the Great Barrier Reef Marine Park, Australia, have yet to institute systematic biotic inventories, despite stated management concerns to "monitor biodiversity."

Geological frame of reference for biodiversity

The term "hotspots," a geological term originally used to describe fixed volcanic sources, has recently gained popularity in the biological sense to indicate areas of high biodiversity. This unfortunate amalgamation of terminology confuses the original use of the term. In this chapter, the term "hotspot" is used in a geological context. Recent efforts to test the hypothesis that geologic hotspots accurately record the passage of lithospheric plates generally have shown it to be valid. If hotspots indeed are fixed with respect to one another over significant periods of geological time, then they would constitute an accurate frame of reference that could be used to determine precise lithospheric plate movement and allow reconstruction of plate positions back into time.

Morgan estimated that hotspots migrate less than 5 mm per year relative to each other, Hotspot trails have been clearly delineated for the time span of zero to 100 million years, thus allowing reconstruction of southwest Pacific tectonic elements This is done by first determining linkages between individual rifted continental fragments and island arcs and then by assigning correct plate motion derived from hotspot models, thus providing a "plate circuit". Such approaches to tectonic plate circuitry allow visualization of the complex movements of plates through space and time.

Yan and Kroenke provide a CD-ROM that contains animated sequences of the southwest Pacific plate movements from zero to 100 million years ago in 0.5-million year increments that provides additional insight on how and when marine organisms may have dispersed or been constrained by various geological events over the last 100 million years. The island of New Guinea, composed of Papua New Guinea (PNG) and Irian Jaya, is located on the leading edge of the

Australian plate that has been moving rapidly northward. The country of PNG comprises a landmass slightly larger than California. The north coast of PNG has been formed by collision events along the north edge of the Australian plate and subsequent "docking" of the east Papuan composite and other terranes in the mid-late Miocene 15-20 million years ago. I suggest that the composite marine fauna of the Madang Lagoon is the result of the accretionary process along the north coast of PNG, involving rapid volcanic uplift as islands, archipelagos, submerged plateaus, and continental terranes are thrust rapidly upward in the collisional process. This process differs from other areas of the circum-Pacific, where terranes are predominantly of oceanic affinity. This "docking" process introduced a number of previously discrete biotic assemblages that then intermingled with established floral and faunal elements.

Reefs in the Madang Lagoon illustrate the importance of taxonomic data in identifying areas of scientific concern. At first glance, the reefs in the Madang Lagoon appear "less spectacular" than their southern counterparts on the Great Barrier Reef. However, taxonomic surveys of marine invertebrates suggest that Madang reefs are some of the most biologically diverse reefs yet documented. Sustaining native levels of biodiversity in these reefs as a potential genetic seed source for other South Pacific reefs is important in the larger context of regional biodiversity.

Marine invertebrates of the Madang Lagoon

Over the past several years, increasing numbers of scientists have focused their research efforts in the Madang Lagoon in an attempt to document the unusual levels of biodiversity found there. The following information summarizes briefly some of the results and trends that have been documented. A more complete discussion can be found in the Proceedings of the seventh International Coral Reef Symposium.

Scleractinian corals. While much attention has to the scleractinian corals of eastern Australia the nature of reef corals from northern New Guinea was not well understood due to lack of distributional data. Potts suggests that the Madang Lagoon may prove to be the single most diverse site in the world for scleractinian corals. Hoeksema's recent treatment of fungiid corals represents the only reef coral family on which detailed distribution and taxonomic data ;it c available for the Indo-Pacific region. According to Hoeksema, northern New Guinea appears to have the highest fungiid biodiversity, with a fauna most similar to that of the Philippines and eastern Indonesia.

Diversity in eastern Australia was the next highest, followed by western Australia and Taiwan, northwest Java, southern Papua New Guinea, and northeast Borneo. Hoeksema stresses that the generic diversity of hermatypic corals in the Indo-Pacific region is quite large, and that lists of genera are not as informative as species diversity and monospecific genera. Because all taxonomic categories above the species level are arbitrary, lists that include species and monospecific genera are more likely to reflect evolutionary history, and thus allow for a more piecise comparison of biodiversity.

Octocorals. Winston estimates that only 50% of the octocoral fauna from the Indo-Pacific region is known at present. Like most other marine invertebrate groups, our basic taxonomic knowledge of the Indo-pacific octocorals is poorly known. A recently described species of octocoral from the Madang Lagoon was unlike any species of the genus previously recorded.

Amphipods. Thomas found that the amphipod fauna from coral reefs of the Madang Lagoon exhibited exceptional levels of species diversity. The amphipod fauna of Madang reefs is a composite, consisting of approximately 180 species, 60% of which are new to science and exhibit multiple biogeographic affinities.

The Madang Lagoon amphipod fauna is taxonomically distinct from other South and Indo-Pacific sites, and the amphipod biodiversity on the north coast of PNG likely exceeds that of any coral reef area studied thus far. However, many coral reef systems in the Indo-Pacific have never been systematically analysed for smaller crustaceans. The author suggests herein that future biodiversity inventories and surveys be undertaken with regard to selective criteria.

Due to the limited dispersion capabilities and habitat specificity of amphipods, amphipods may be of use in biogeography and environmental monitoring in coral reef systems.

Crinoids: Messing reported 39 species ofcomatulid crinoids from the Madang Lagoon and, with limited sampling, found the crinoid fauna of the Madang Lagoon comparable to other more intensively studied sites such as Lizard Island and Davies Reef, and Palau.

Gastropod molluscs: Working in the Madang Lagoon, Gosliner found that the north coast of PNG supports a more diverse fauna of opistobranch gastropods than has been reported from any single geographical area studied thus far. The next richest tropical site is Guam, followed by Hawaii, the Caribbean, and Japan. Gosliner's faunal records are significant because of intensive field efforts in numerous tropical localities using snorkeling and

SCUBA that enable comparative studies. Other areas that are known or suspected to house high diversity have not been studied adequately to allow comparisons of opistobranch biodiversity.

Biota and Habitats: Kristian Fauchald reported that the polychaete fauna of the Madang Lagoon exceeded that of any area yet sampled. Clyde Roper and Mike Sweeney reported similar findings for cepahlopod molluscs. In a preliminary survey of the marine algae of the Madang Lagoon, Mark and Diane Littler reported that not only were there more species of algae collected, but the number of undescribed species surpassed that of any region previously sampled. The geological history that may have contributed to the extraordinary levels of marine invertebrates in the shallow waters of the north cost of PNG also may have influenced the fauna in deeper waters of the region.

Studying a collection of deep sea crustaceans from the Bismarck Sea region Austin Williams of the National Marine Fisheries Service Systematics Laboratory reports unusual levels and types of biodiversity. More investigation of the deep sea component of this region is warranted in light of this preliminary information.

Selecting coral reef sites by attributes

Most current approaches to protecting biodiversity place emphasis on areas or species of special human interest and natural beauty with areas with novel biological features. Therefore, highest value is placed on unique and unusual biota, with little regard to the nature and quality of biodiversity on a larger scale. It is imperative to develop priorities using the geological past, the nature of the ecosystem, the characteristics of the region and the available specialists.

12

Tropical Sustainable Development

Sustainable development is tied indirectly to biodiversity through the need to maintain overall environmental values. Its purpose is to sustain both man and nature, and thus it is directly related to conservation and the wise use of resources, including biodiversity. Sustainable development commonly is defined as the process of meeting the needs of the present generation without compromising the ability of future generations to meet their needs.

This concept arose in the late 1980s as an approach to balance economics and environment. Sustainable development embodies several advancements over other models that have characterized the development process. One advancement is drawing distinctions between the growth component and the efficiency component of development. The growth component refers to an increase in size or dimension of an economy.

Assuming that an economy draws on a finite resource base, growth cannot continuous indefinitely, i.e.., it cannot he sustained. The efficiency component refers to the "realization of potential," which is a more qualitative sense of development that may be able to be sustained on a finite resource base. Thus, the recognition of the different components of development can lead to more effective policies for resource use that focus on efficiency rather than growth. These ideas, along with recognition of the importance of future generations and the value of the environment, make sustainable development an attractive philosophy for planning.

Table. 1. Techniques for mitigating deforestation and maintaining tropical forest habitat

Strategy	*Mechanism*	*Examples*
Preservation	Deforestation is eliminated as a sanctuaries Biosphere reserve	National parks Community land-use option
Conversion from extensive to intensive forms agriculture	Less land is used to produce the same amount of yield, thus freeing other land for forest preservation	Polycropping Intercropping Fertilizationof Agroforestry Intensive pasture
Extraction of products and Harvest of non-market values	Forest is because of values from maintained timber products	Selective logging Ecotourism demonstrated existing forest
Reforestation	Restored forest directly replaces deforested land	Managed succe-ssion Tree planting

In the tropics, conservation has focused on controlling the deforestation process. All of the strategies listed in Table 1 can be thought of as forms of sustainable development, but the extraction of products and values from existing forest may be the approach that most directly links biodiversity to sustainable development. The idea is that forests or other habitats can be maintained if they can be shown to provide direct value to people, especially rural people who coexist with the forests. Direct market values of forests are derived from the ecosystem as a whole for ecotourism and from harvest of individual species for timber or nontimber resources. Of course the value of logging or the harvest, of trees for timber is well known but assessments of value of nontimber, products are very recent.

The much-cited paper by Peters et al. led the way in showing that harvest of nontimber products from intact forests can generate more revenue than other forms of forest land-use that are more destructive, such as logging or conversion to pasture or tree plantations. Although the estimate of the value of the forest given by Peters et al. may not be representative of tropical forests in general, their study has led to a new vision of rain forest economics. Harvest of medicinal plants is another important example of this approach, but many other nontimber resources potentially can be sustainably harvested from intact tropical forests.

The Mckelvey box classification of boidiversity

In viewing extraction as a form of sustainable development biodiversity is a resource to be Utilized by people. In fact, each species is a potential resource. To clarify the role of species in sustainable development, an analogy can be made with a classification used for mineral deposits. The McKelvey box separates mineral deposits based on the knowledge of their existence and their economic concentration.

The term "reserves" is applied only to those deposits that have been identified and are in sufficiently high concentrations to be mined economically. All other deposits are termed "resources." This classification provides a status assessment or inventory that is useful for decision-makers. The numbers shown in these McKelvey boxes are tentative, but several patterns are evident.

First, there are relatively species that currently and directly are utilized by people and therefore fall into the 'reserves" category. Also, there is a high proportion of undiscovered species especially insects. Many and probably most of these undiscovered species occur in tropical forests, most of which are undergoing deforestation.

Although these undiscovered species present conceptual problems, their loss to extinction certainly would be significant. It is important to recognize that the McKelvey box represents a single time frame and that over time the numbers inside the box change. As, taxonomic explo-ration continues, the number of undiscovered species will go down. As technology advances, the number of economically useful species will increase. New mechanisms are developing to stimulate these changes.

Biodiversity prospecting is being, advocated, and mechanisms for distributing income that is generated from biodiversity are being developed. These mechanisms include the work of national institutions such as the Institute National de Biodiversidad de Costa Rica (INBio)and, at the international scale, the Biodiversity Treaty.

Impediments to sustainable development

The idea of sustainable development is attractive but

there are a number of problems or impediments in practice. This should riot be surprising, since the system of man and nature that is to be sustained is complex. Some of the problems with sustainable development are described below. To a greater or lesser extent, these and other problems must be addressed for success to be achieved. Overharvesting or overuse is a critical issue in most resource systems. Because income often is perceived to be directly proportional to harvest, users of resources easily can be enticed into overharvest.

Thus, attention must be given to harvest and use schedules. This requires the best scientific knowledge, but it also requires that users of resources face the issue of limitation. In some cases, short-term sacrifices in harvest may be needed to achieve long-term benefits of sustainability. In this regard, management of people is just as important as management of resources in sustainable development. One approach to supplement income of rural people is to develop new markets, either for existing products or products of new resource-use systems, such as harvest of nontimber products from forests.

The hope is that, by providing additional income derived from forests, there will be less incentive to cut the forests. An interesting example of this strategy is the opening of markets *in* the developed countries for tropical forest products. This is difficult to achieve, but strategies are being developed. Existing political economics in the tropics can promote environmental impacts. The recognition of this problem has led to an exciting new approach, termed "political ecology". This approach goes beyond combining economic and ecological perspectives to add the political context within which land-use decision are made.

Political ecology is helping to identify the complexity of tropical problems, but-new political economics are needed. Accounting systems are being developed that may fill these needs, but they are difficult to implement. Areas that are undergoing economic upheaval, such as Brazil and the Eastern European countries, may lie the most likely candidates for the implementation of these new approaches to accounting and valuation.

The typical pattern of land-holdings for a tropical country is where a large portion of land is controlled by relatively few wealthy people and the remaining small portion of land is divided among a large number of poor people. This uneven distribution leads to social problem, that can inhibit wise land-use strategies and encourage deforestation. Approaches for redistributing wealth, such as the development of new markets for rural people described earlier, may help mitigate this problem.

The concept of extractive reserves was developed in part to deal with this problem. Rural people would have access to these lairds to generate income through harvest of nontimber products. However, controversy exists as to whether or not sufficient income can be derived from extractive reserves, at least under conditions in the Amazon. The problem with government subsidy of bad land-use has been demonstrated by analyses given in Repetto and Repetto and Gillis.

These authors provide many examples, perhaps the best known of which is tile tax incentive program that was provided to ranchers in the Amazon during the 1970s and 1980s. This program led to much deforestation and subsequent declines in productivity of pastures. Political backlash call occur when one country interferes with the sovereignty of another. Several examples of this phenomenon are known From Brazil whose government has been criticized for deforestation by conservation groups in developed countries. Stimulated by a World Bank publication, a major controversy arose concerning rates of deforestation in the Amazon. As a consequence, the President of Brazil even stopped accepting financial support from foreign conservation organizations because he feared that they were having too much influence on Brazilian land-use decisions. This position was soon reversed, but it illustrates the potential magnitude of political backlash.

Other Amazonian examples of backlash have involved a benefit rock concert and the advocacy actions of a foreign anthropologist. War and violent conflict can severely inhibit conservation and sustainable development. During such times, priorities change and environmental values frequently are overlooked as conflicts are resolved.

History lessons for tropical sustainable development

One way to gain perspective on the problems associated with sustainable development is to look to history. Comparative analysis of these examples potentially can provide inspiration and insight for present-day issues. Historical fallures or problems can be lessons to learn from, and historical successes can be models to emulate. Thus, using Deevey's metaphor for historical ecology we "coax history to conduct experiments" in sustainable development with 10 examples.

Henry Ford's sustainable villages

Henry Ford was a dynamic man who revolutionized America through his contributions to the auto industry in the early 1900s. He developed an effective line of automobiles and an industry that could mass-produce them. In developing this industry, he also pursued related social agendas that included worker benefits, international relations, and the state of agriculture. One of his agendas was the decentralization of his industry and the back-to-the farm movement. This philosophy became embodied in his "village industry" net work in southeastern Michigan. In each small town, he built a factory for automobile parts, usually manned by part-time farmers and supported by a hydroelectric power plant.

Each community had an element of self-sufficiency, but they also were connected to each other through the parts of the automobile that the produced. This was a type of utopian plan that was fairly successful until the region became more diversified with the growth of the country. Perhaps Ford's village industry philosophy Could be a model for tropical landscape with sustainable mixtures of wilderness, agriculture, and towns.

The American Civil War and the North-South Dichotomy

The Civil War in the United States was the most costly war in American history in terms of loss of life. It originated from a North-South dichotomy that is reminiscent of the current global dichotomy between developed and less-developed countries. The American North-South dichotomy arose early in colonial times from differences in cultural background of the settlers and differences in land-use as determined by geography and climate. The dichotomy solidified after the revolutionary war, with the North becoming primarily industrial and commercial and the South remaining primarily agrarian. At the beginning of the Civil War, these differences were dramatic.

The North had live times as many factories and more than double the Population of the South, which was dominated by monoculture crops such as cotton, tobacco, and rice, as well as pastures for beef cattle production. Slavery was common to both the North and the South before the American Revolution and was not the primary factor that originally distinguished the two regions. The end of the Civil War was followed by the Reconstruction Era, which lasted 10-15 years and consisted of political attempts to manage the process of reuniting the divided country In general Reconstruction was dominated by corrupt politicians and shrewd businessmen who failed to generate economic and social development.

The obvious lesson of the American Civil War is that dichotomies, "such as that between the North and South, which are characterized- by extremely unequal distributions of wealth and industrial develop-ment and strong relations of dependency, are not stable and can result in violent conflict.

The Reconstruction Era also may provide lessons on how not to stimulate development in a depressed region. There are similarities with recent development efforts in Africa, which have been described by the metaphor of "tropical gangsters". Thus, some nations in the tropics now are facing some of the same problems of development that the South had to face after the Civil War.

Tropical partnerships for transportation and development

An important aspect of tropical development has involved the actions of large foreign companies. The American-owned fruit companies of Central America are good examples: Standard Fruit and United Fruit. These companies began operations in the Atlantic lowlands of Central America in the late 1800s with the primary focus of production being bananas. Towards this objective, thousands of hectares of rain forest were converted to banana plantations in an export business which was established to supply the demands of a rapidly growing market in the United States. As time passed, the negative aspects of this form of land-use became obvious: loss of watershed protection, reduction in diversity local economic dependence on a monoculture, exportation of the crop, and degradation of water quality. Certainly these are social and environmental impacts that perhaps could and should have been avoided.

However, at least in there also were positive aspects of the presence of foreign companies in the tropics including improvements in health care and infrastructure. In particular, the construction of roads and railroads was a benefit in the sense that not have been available for these developments without foreign input. Often the transportation networks were designed to meet the needs of the company, i.e., movement of bananas to coastal ports, but they also were used for the benefit of local populations.

Transportation infrastructure, especially road has been recognized as leading to deforestation and nonoptimal land-use by opening terrain that previously was not available to land-poor farmers. The tropical road has become "the symbol of modern deforestation". Perhaps this negative view needs to be balanced by the positive aspects of transportation networks.

One example of the positive aspect has been noted by Clay and Clement "The development of markets for sustainably harvested commodities and the destruction of tire rainforests both depend, ironically, on the, same thing: improved transport systems ... Forest residents have long realized that roads are both their salvation and their demise." Although alternative transportations systems need to be researched, roads will continue to play an important role in tropical development. In this regard, partnerships between conservationists and foreign companies, such as the old United Fruit Company, may provide the means for this planning.

The Columbian exchange and marketing of nontimber products.

Although Christopher Columbus may have been disappointed that his voyages did not discover the trade route to Asia, the "Columbian Exchange" of materials, species, and cultures that he initiated was dramatic. One of these exchanges may represent a model for current efforts to develop markets for rain forest products.

The New World provided many species of food for the Old World, including maize, potatoes, cacao, and others. Some of these were similar to species used for food in the Old World and diffused easily. Others such as potatoes and tomatoes, were unlike anything in the Old World and thus, took longer to be accepted. Davidson describes some reasons why people were reluctant to take up certain New World foods and why other foods were accepted quickly.

The process of acceptance of a foreign food item that occurred with these species from the New World may provide ideas and interesting contrasts with efforts being made to market rain, forest products in the United States and Europe. Clay discussed this new marketing- challenge, drawing heavily on his experience with the organization called Cultural Survival. This effort requires information on biology, ethnology, food science, international trade, and business.

Several nongovernmental conservation organizations are working on these exchanges, which involve paying producers a fair world price or sending a portion of the profits back to tile tropical cultures that produced the item, as with other forms of biodiversity prospecting. The marketing of rain forest products is not limited to food items, and the New World to Old World "Columbian Exchanges" also may provide useful insights for facilitating other efforts in present-day marketing of products related to conservation.

Amazonia and the Louisiana purchase

The vastness of the Amazon basin long has been a stimulus to the imagination of people who wish to exploit its resources. This has led to explorations from the time of Orellana to recent times. These explorations resulted in both a romantic literature and assessments of resources that have been used as a basis for public policies. Unfortunately, the images of the Amazon have not always been accurate and have lead to unrealistic expectations. Jordan describes one example in relation to agricultural potentials, which never were realized.

The lesson here is that images are an important part of the development process, but that they also call be detrimental to the extent that they mislead development careful analysis of historical examples, such a, the development of the American northwest, can provide useful information for the development of tropical areas Such as the Amazon.

Export agriculture in central America

Central America's economy has been characterized by a reliance on export agriculture. Originally, the focus was the "dessert economics" or luxury, crops of coffee, sugar, and bananas. After World War II, cotton and beef gained in production. These exports primarily have gone to Europe and the United States, as in Central America. The resulting relationship of dependency has negative effects on many aspects of the Central American economy, environment, and society. A similar situation has characterized the southern region of the United States. In fact, Persky use the Latin American example as a model for southern economic history. A strong dependence arose, with the American planters exchanging tobacco for clothes, farm machinery, and other supplies from English merchants. This was a classic case of exchange of raw materials for manufactured goods with a trade deficit. Accordingly, the planters fell seriously into debt over time. Many went bankrupt and were thrown into debtors' prison, but a few were able to switch to more balanced agricultural production that served a domestic market.

The best example was George Washington, who gradually switched from production of tobacco to wheat and corn, which he exchanged with merchants in nearby Philadelphia for American-made goods. Other examples are described by Clemens for Maryland's Eastern Shore. Thus, some of the colonial planters were able to diversify their production and reach new markets. This is a strategy that has been recommended for Central America. The transition will be difficult, but helpful insight might come from analysis of the colonial Americans who were able to escape dependence on England some 200 years ago.

Civilian Conservation Corps and Tijuca forest

The Civilian Conservation Corps (CCC) probably was the most massive program of human and natural resource management in American history. From 1933- 1937, more than a million young men were employed by the federal government to work on a variety of natural resource projects. The CCC program was devised by Franklin D. Roosevelt as part of the New Deal, primarily as a way to increase employment during the Depression. The program was restricted to unemployed and unmarried young men between 18-25 years of age who signed up for tours of between

6 months and 2 years. During work tours, men were clothed, housed, fed, and paid a modest wage. Men lived in simple, rustic camps located near work projects, which included such activities as forestry, prevention of soil erosion, and flood control.

The program involved a rapid, large-scale mobilization of both manpower and facilities and was accomplished by incorporating the War Department in a supervisory role. The program also had the goal of training the men so that they could find employment after they left the Corps. Thus, the CCC was both an ambitious social program that provided employment and training to primarily unskilled men and a major capital works program that resulted in the following significant accomplishments, which occurred after just the first year: ". . . construction of 25,000 miles of truck trails; 15,000 miles of telephone lines: 420,000 erosion check dams; disease and insect control on 3 million acres of forest; 98 million seedlings planted; forest stand improvement on a million acres; and 687,000 man-days of fire fighting". Another example of this type of large-scale deployment of manpower occurred with the reforestation of Tijuca Forest in Rio dc Janeiro during the late 1800s.

The Emperor of Brazil had the coffee plantations converted to forests in order to save the water supply of the city. It was a massive program, organized by the army and carried out by slaves, that involved the planting of more that 100,000 trees. Today, Tijuca Forest is a national park with a magnificent rain Forest that is surrounded by the city of Rio de Janeiro. In both historical cases described above, a large-scale, labor-intensive public works program in natural resource management was conducted in response to a crisis.

13

Information Management for Biodiversity

In European have worked together to build a marvelous weather information infrastructure. Consider how pervasive weather information is. We hear about the weather on the radio, see it in the newspaper, and now television provi les spectacular satellite images and movies, Doppler radar displays, animated graphics, and lively human commentary to explain current conditions and forecasts. Weather is part of every local news show, and in the United States we have the Weather Channel on cable.

We have spent billions of dollars oil capabilities to gather weather data and generate and disseminate forecasts. Weather satellites and information networks provide global coverage. By most reckoning, thousands of lives have been saved by improved forecasting and warning capabilities, and large economic losses have been avoided through timely responses to better predictions. Finally, consider sports, another area of pervasive information lows. For most of us, professional sporting events lack the immediate, direct impacts oil our lives that financial outcomes and weather events have. Nonetheless, millions of people are keenly interested in sports, and another elaborate information infrastructure has been built to satisfy that interest. We can speculate about which came first: the overwhelming interest in professional sports, or the vast information infrastructure that stimulates and satisfies that interest.

Sports information captures large sections of the printed media and is, inescapable in broadcast media. Sports has its own publications and radio and television networks. And the wide availability of sports information influences behavior. Biodiversity, which affects our lives in the most fundamental ways possible. People are extraordinarily interested in the environment and biodiversity. Recent polls indicate that approximately 80% of Americans call themselves "environmentalists." That interest extends to the natural environment.

More Americans visit museums and zoos every year than attend professional sporting events. Tens of millions of Americans enjoy various activities outdoors in ecosystems: hiking, birding, fishing, hunting, camping, and so on. When they are exposed to good information about biodiversity, young children show great interest-coin parable to the levels of interest that they later demonstrate for sports. Not only is interest keen in biodiversity and the environment, but, unlike the weather, humans can do something to change the situation; and we do a great deal.

We eliminate species like the ivory-billed woodpecker and passenger pigeon. We introduce species like the gypsy moth and zebra mussel. We alter and destroy habitat. Sometimes we restore ecosystems and reintroduce native species that had been extirpated. In short, unlike the weather, we can do much to change the environment for the better or worse, Each of us takes actions that to various degrees help or harm.

The present administration of the United States is committed to an ecosystem management approach to the environment and the economy. The concept of ecosystem management explicitly recognizes the linkages between ecology and economy and rejects the assumption that there is an inherent conflict between jobs and the environment.

There are strong parallels between financial management and conservation of biodiversity. If we do not know what our ecological assets are, we are poorly equipped to safeguard those assets through wise use and management. If we "spend" too much of the natural production and capital of an ecosystem, we see symptoms of impending bankruptcy, e.g., the collapse of fisheries endangered species, and loss of jobs.

Managing ecosystems by deficit-spending results in what Secretary of the U.S. Department of the Interior, Bruce Babbitt, calls "train wrecks," with examples like the forest and fish crises in the Pacific Northwest. Information is equally critical to financial and ecosystem management. Without information, we cannot manage finances at the personal, corporate, or governmental levels, without information, we cannot manage personal "back-yard," local, or regional ecosystems so as to conserve biodiversity.

The U.S. National Information Infrastructure

The United States is building its national information infrastructure, (NII) a national system of information superhighways. It is a given that the NII will offer all the financial weather, and sports information that we could possibly want, in fact more than we can possibly absorb. Through a National Biodiversity Information Centre (NBIC), the NII also could be rich source of information on biodiversity of the public.

Regrettably, in the United States, information on biodiversity receives a minuscule fraction of the attention that is paid to finance, weather, and sports. Lack of information is the silent partner to the more prominent agents of environmental destruction. If the destruction happens gradually enough and invisibly enough, Without galvanizing incidents like the catastrophes of Bhopal and the Exxon Valdez, by the time a threshold is crossed and the train wreck it is too late. We need to make the gradual, seemingly invisible changes visible to people. Financiers and meteorologists know this lesson. Information is the key to an effective strategy of empowerment for the environment. If we were to build a robust environmental information infrastructure, comparable in capabilities to those of finance, weather, and sports, we could provide people with environmental information to empower them. This kind of empowerment would work the same way that financial information empowers us to make informed choices in our financial management, weather forecasts empower us to prepare for the weather, and sports coverage empowers us to feel a strong emotional involvement with sporting events.

Empowerment through environmental information requires a new way of thinking. Our robust weather information infrastructure follows an implicit "observe-forecast-warn-react" model. This works quite well for the weather, since we cannot change it and can only react to it. But this model is too limited for the environment, where the relationship is more complex—the environment changes us, but we also change the environment.

We need to follow a "feedback" model that acknowledges the feedbacks between humans, information, and our environment. Consider the well-known impact that information has on the stock market. Responses to information about actual events and rumors can cause waves of buying or selling, moving prices up and down. Analysts know that there is a highly dynamic feedback process between information and markets. If we adopt this feedback model, we see that information is the fundamental mechanism that influences human behavior toward the environment.

Initiatives for the conservation of biodiversity call succeed only to the extent that we put information to work as a stratagem, resource, and tool. The weather model has been the dominant model in thinking about the application of the information infrastructure to the environment. If the weather model is the controlling paradigm, then it seems that there is not much to observe, forecast, warn of and react to, because most environmental changes take place, seemingly invisibly, on a longer time scale than weather events, which happen on scales of minutes to a few days and pose immediate and obvious threats to life and property.

On the other hand, if the feedback model is the controlling paradigm, it is imperative that we create an "environmental channel" to provide people with the information they need for empowerment and to create feedback loops that change behavior and result in positive changes for the environment. Key components of a vision that is both possible and necessary for an effective environmental information infrastructure have been demonstrated technically in other contexts. Also, the Environmental Resources Information Network (ERIN) in Australia, the Right-to-Know Network (RTKNet) in the United States, and other advanced practitioners already have demonstrated some of these capabilities in the environmental arena.

An environmental channel will offer six major capabilities. First allow its users to explore past, present, and predicted future environmental conditions at local, state, regional, national, and global scales, using ecosystem or political boundaries. Second, it will provide capabilities for modeling, visualization, and simulation to explore alternative future scenarios. Third, it will provide information about biodiversity and other natural resources in local areas and about stresses on those resources. Fourth, it will provide, information oriented toward solutions that will empower individuals to identify what can be done to minimize harm and improve the environment. Fifth, it will allow individuals to "publish" observations, ideas, and questions about the environment. Sixth, you will be able to find information, other resources, and collaborators for environmental efforts.

International models

ERIN

The Australian Environmental Resources Information Network (ERIN), a unit of the Department of Environment, Sport, and Territories, provides an excellent model for the sophisticated application of information tools, technologies, and sustainable development. ERIN demonstrates attributes of the needed capabilities of an environmental channel and of a biodiversity information center. ERIN takes a distributed network approach, in which it identifies the custodians of high-priority data, such as standard spatial data sets, and works with those custodians to make their data available over the information network. The agencies retain custodial responsibilities. ERIN seeks to ensure that primary data are available, so that users are free to make their own interpretations rather than being, forced to rely on derivative data such as classification schemes for land-use or land-cover.

ERIN operates a World Wide Web server, "Australian Environment On line," that is accessible internationally over Internet through Web browser soft ware like the popular Netscape and Mosaic packages. Its uniform resource locator (URL) is http: // www.bdt.org.br/bin21/bin21.html. The ERIN web server demonstrates advanced information technology that brings together a wide range of environmental information and presents it to users, along with tools for spatial display of geographical information, modeling, and viewing.

BIN21

The Biodiversity Information Network (BIN21) initiative, managed by the Base de Dados Tropical in Brazil, was created to provide informational support that would further the purposes of the Convention on Biological Diversity and Agenda 2 1. BIN21 is developing a distributed, international network of sources of information to support research on biodiversity and conservation. Like ERIN BIN21 now operates a World Wide Web server (URL http://www.ftpt.br). BIN2 is working to create International "nodes" on the Internet and foster international exchange of information and collaboration.

WCMC

The World Conservation Monitoring Centre (WCMC) offers yet another useful model. WCMC's Web server is found at URL http://www.weme.org.uk/. Also , WCMC operates the pilot clearing House Mechanism for the Convention of Biological Diversity at URL http://www.wcmc.org.uk/~chm..

U.S. National Biodiversity Information Center

The concept of a focal point for efforts to understand and sustainably man age biodiversity in the United

States has been discussed for over 20 years. 12, 1974. The Nature Conservancy pioneered the first of its State Natural Heritage Inventories in South Carolina. Natural Heritage Data Centers now exist in almost every state of the United States, using methodology that has been improved continuously.

In 1986, a report of the Association of Systematics Collections (ASC), Foundations for a National Biological Survey, and a background paper of the Office of Technology Assessment (OTA), Assessing Biodiversity in the United States: Data Considerations, brought increased attention to this area. In 1987, the OTA made a modest recommendation for "a small clearinghouse for biological data" in its report, *Technologies to Maintain Biological Diversity.*

More recent examples of the continuing interest in the management of information about biodiversity include the 1991 policy dialogue report of the Keystone Center, Biological Diversity on Federal Lands, the 1993 report to the Smithsonian Institution on the National Center for, Biodiversity by the ASC, and the 1993 report from the Council on environmental Quality, Incorp-orating Biodiversity Considerations Into Environment Impact Analysis Under the National Environment Policy Act. The Discussion in these reports dealt with a board carry of activities designed to improve the management of biodiversity, including regulation, conduct of government activities, and improvements in the amount and availability of information.

A Proposal for a National Institute for the Environment, published by the Committee for the National Institute for the Environment, recommended that the proposed agency include a National Library for the Environment with modern information services and electronic technologies. Both the 1993 report from the National Research Council. A Biological

Survey for the Nation, and the 1994 Systematics Agenda 2000: Charting the Biosphere cited decentralized computer networks of databases as absolutely central to the study and preservation of biodiversity.

From 1987-1992, several Congress considered bills that would have instituted various regulatory programs, requires consideration of biodiversity in environmental impact statements under the National Environmental Policy Act (NEPA), and Established a national center of biodiversity. The Bush administration resisted these efforts, primarily due to concerns over NEPA and other regulatory and quasiregulatory and quasiregulatory provisions.

In these bills, the concept of a center was quite broad. A center would be the home of fairly ambitious efforts to (1) locate information that is available on the nation's biological resources: (2) assess the completeness of that base of in formation; (3) catalyze and, in some cases, actually fund efforts to fill gaps in information; (4) serve as a clearinghouse, or "database of data bases," that would facilitate improved access to information by users; and (5) generate its own products, such as maps, flora, and other types of publications and media.

At the Earth Summit in Rio, in -June 1992, file Bush administration embraced the idea of a national center of' biodiversity, based on a more limited conception of its scope and role. A center was envisioned to include three general functions: (1) developing methods to access widely scattered information on biodiversity, (2) assessing the state of knowledge and identifying gaps, and (3) providing leadership in understanding and communicating about biodiversity. Plans were developed to establish by Executive Order a center at the Smithsonian

Institution, but the order was not signed before President Clinton took office.

A Biological Survey for *the Nation* recommended that "Under the leadership of the NBS, the National Partnership for Biological Survey should develop a National Biotic Resource Information System . . . (The NBS) should also participate in interagency initiatives to coordinate collection and management of biodiversity data by the federal government.

Recent Planning

With the establishment of a National Biological Survey, the Environmental Protection Agency provided funds to the Smithsonian Institution to lead an interagency planning study, beginning in the fall of 1993, to clarify the concept a National Biodiversity Information Center (NBIC). A small ad hoc working group of staff that represented several federal agencies developed background materials to prepare for the establishment of a center. The materials included a "strawman" description of what might be included in a National Biodiversity Information Center and questions to be addressed in the planning process.

The Office of Management and Budget shared the results of this working group with the assistant secretaries of over 10 federal agencies involved ill biodiversity research at a meeting in January, 1994. After receiving advice oil policy from the assistant secretaries, the Smithsonian Institution began the next phase of planning with representatives of federal agencies, state agencies, academia, museums, nongovernmental organizations, and industry.

The Board was responsible for setting the specific charge for and overseeing the work of the Drafting Committee, which was chaired by Dr. Robert S. Hoffmann, Assistant Secretary for Science at the

Smithsonian Institution, and included members from the full spectrum of stakeholder organizations. The Drafting Committee produced a draft report that was taken to the meeting of the Board on November 15 for adoption. The report provided both a conceptual framework and a detailed account of the structure, location, and institutional setting of the Center and its principal tasks. Relationships between the Center and other organizations also were addressed.

Changes to the draft report, suggested at the Board meeting of 15, were incorporated into the final consensus paper published in December 1994, The full text of the final report, minutes of the two meetings of the Advisory Planning Board, and a list of Board members can be obtained on the 'World Wide Web through the Biodiversity and Ecosystems Network (BENE), URL http://straylight.tamu,edu/bene/nbic/nbic.html;via anonymous FTP from keck.tamu.edu/pub/bene/bene-texts; or from the NBIC area on the Smithsonian Institution's National Museum of Natural History Web and Gopher servers. The principal concepts coming from this report are presented below.

Mission and objectives

The idea of the National Biodiversity Information Center is simple: it provides a focal point where the many parties that generate, manage, or use data on biological resources can collaborate and make decisions leading to broader access to that information. The Center will direct users to sources of the data they seek, while working with funding agencies to encourage development of tools and strategies to make data more accessible. The Center will not duplicate existing, databases or information, but will provide directory services for the large array of information that is available. It also will identify gaps where new databases are needed, and assist the development, transfer, and application of new technologies. Further, the Center could coordinate access to data outside the usual of the biological sciences.

NBIC's mission will be to provide leadership and a neutral venue in which to facilitate collaborative discussions about the availabtity of data and information on biodiversity. This will be accomplished through an interactive computer system that uses metadata on geographic location, species, ecosystem, or other keywords to sort, aggregate, or integrate data sets, identify gaps in existing data knowledge, and provide a forum for collaborative approaches to issues in biodiversity information.

Guiding principles

The Center must be responsive to the needs of users, providing both data and information services that are tailored for different audiences. NBIC also must be responsive to the needs of providers, and must offer incentives and encouragement for them to provide their data through NBIC. The Center will facilitate the development of standards for metadata minimum criteria for the documentation and format of data and the establishment and provision of protocols for collecting and reporting data. Guidance on appropriate uses of data or information also will be provided. NBIC will facilitate the improvement in the quality of data sets with a feedback system that allows comments on the quality and utility of data.

Custody of data will reside largely with primary collectors And producers of data, and users will be referred to Original sources of data. Therefore, data holdings by NBIC will be reduced. NBIC will use appropriate technologies for the integration and analysis of information and will promote tile adoption and use of appropriate standards of information.

Structure and location

The Center will have a distributed structure that will function on a democratic, consensus- building, and partnership approach. NBIC will serve as a convener, facilitator, and host. Center experts will move discussions along and involve key constituencies. An Advisory or Governing Board from the broad community of contributors and users will provide general direction. A Technical Advisory Committee will ensure that NBIC receives computer and technical advice, NBIC must establish partnerships with the other organizations, such as the NBS, whose activities include collection of data and information and assessment of biodiversity issues.

NBIC's location should be decided by open competition, with a review after 3 years and recompetition every 5 years if so recommended by the review. NBIC should be designed to be moved if needed. Individual institutions consortia of institutions should be invited to apply for selection as the Center host.

Desirable characteristics of the host institution should include strong support, for computational and information management services; a creative and active gram in biological sciences, especially involving the use of computers diversity information management; broadly based expertise and strong systematics, ecological research, and information management of collection understanding of modern and historical North American collections history of service and support to users; reasonable access to national and international transportation; and comfortable, modern facilities for conferences and the staff of the center.

The National Biodiversity Information Center (NBIC), the National Biological Service (NBS), and the National Biological Information Infrastructure (NBII)

Both the NBS and the report of the National Research Council, A Biological *Survey for the Nation,* conceived of the establishment of a highly distributed information resource that would involve participants from federal, state, private, academic, and other institutions at scales ranging from local to national and global. Participants in this National Biological Information Infrastructure (NBII) would work to develop tools, strategies, and collaborative activities that will bring coherence to the collection, storage, and management of data and increase access to biological information.

The newly created NBS has as a significant component of its mission the augmentation of access to biological information, relying upon a "distributed federation of databases" in accordance with the recommendations of the National Research Council. To achieve this goal, NBS plans to gather and make available information on sources of biological information encourage the development of standards for metadata, and make investments in both its own and its partners' hardware and software capabilities to move toward a highly distributed communications and access environment.

These activities and objectives closely complement those of the NBIC. The Center will assume the clearinghouse function of the NBII as envisioned by the NBS. NBS views the Center as a powerful tool for achieving key NBS objectives. The development of the "National Partnership for Biological Survey" requires that a wide range of partners meet regularly to work toward common strategies for achieving specific partnership goals. NBIC provides the "neutral forum" around which those common strategies can be developed. The relationship between NBS and NBIC will evolve, with a shared goal of improving the accessibility of data and information on biodiversity. The Center by representing not solely a single government agency or the federal government but the entire biological information community in the United States—can provide important international linkages and like an important statement regarding the need for broad collaboration.

14

Farming with Biodiversity

Agricultural progress seems even more paradoxical when we consider biodiversity use in agriculture. Several thousand plant and animal species are known to have served as food in human history. The displacement of traditional crop varieties by newly-developed high-yielding varieties has been a major factor in this development. FAO estimates that in the next 10 years 90% of the existing genetic diversity within the main crops is at serious risk of disappearing. The vast majority of the literature dealing with biodiversity use in agriculture is characterized by a symptomatic approach, and centers on providing data that prove the continuous reduction of both inter and intra-species biodiversity used in food production, on its associated risks for the sustainability of our food supply, and on alternative farming techniques to halt the erosion of biodiversity use in agriculture.

A discussion of the driving forces that have generated the reduction of biodiversity use in agriculture is generally missing. Put another way, the detection of symptoms is directly coupled, to a prescription of remedies without any serious attempts to diagnose the disease. Perhaps this is because a thorough diagnosis would necessarily raise some politically unpleasant issues. The decreasing energy output/input ratio in agriculture means that the food security of humankind is increasingly dependent on rapidly dwindling oil reserves, and on an increasing environmental impact per unit of crop output. The dramatic reduction of biodiversity worldwide spells serious problem for the maintenance of ecological, equilibria at the planet level. That is, the path of technological in food production chosen right now by humankind is not sustainable. Not only are these unpopular issues to discuss, but it is becoming evident that most of the undesirable side-effects leading to unsustainability are difficult, if riot impossible, to incorporate in conventional sets of measures and indicators used to assess risks, costs and benefits of technological change. In spite of these difficulties, it is crucial to understand the driving forces that are shaping current trends of technological development in Otherwise, we seriously risk caveats in discussions of future agricultural policies. It is common knowledge that traditional techniques achieved high output/input energy ratios and used much more biodiversity than modern high-input agriculture.

Individual scientific disciplines that study changes in socioeconomic systems tend to focus on the effect of one factor at the time. Although this approach provides a high resolution for the factor in question, it does not shed much light on the bigger picture. In this respect the agricultural sciences are no exception.

Ecologists tend to focus on the on-site and off-site effects of agriculture on the environment, without considering agriculture as a livelihood. Economists are concerned with the economic return of the labour and capital resources invested in the farming process, and tend to more the ecological impacts. Agronomists occupy themselves basically with the physical performance of the production system. Rural sociologists and anthropologists are mainly interested in the effects of technological changes on the lives of individual farmers and the identity of rural communities. This specialization or *reductionism* of established scientific disciplines is at times necessary, but often creates confusion for decision-makers that have to decide about the future directions of technical development in agriculture.

All of the above-mentioned aspects of agriculture must be addressed, since each one is fundamental for the sustainability of agriculture. The problem lies in the fact that a detailed discussion of any one of these aspects becomes useless if not put into relation to the functioning of the system as a For this reason, it is of utmost importance to put the different disciplines in perspective within a common frame of analysis.

The existence of different perspectives for assessment of stistainability of a agricultural development has been pointed out by many workers. In spite of minor differences in definitions and emphasis among the different authors, there is substantial agreement that at least three fundamental perspectives should be considered when attempting to define sustainable agriculture:

- The ecological view: agricultural techniques must be environmentally sound;
- The economic view: agricultural techniques must be economically viable;
- The social view: agricultural techniques must be acceptable to farmers and society given culture, ethics and religion.

Assessing only agronomic performance or ecological

compatibility of various agricultural techniques is not sufficient to evaluate their sustainability. We need a parallel check on economic viabilily and social acceptability. 1 believe that it is thus safe to conclude that we need to focus our efforts on the development of frameworks or models of analysis that establish bridges among the different perspectives. The farming system as a sub-system of society: constraints imposed by population density, income level and labour time allocation in society. In order to verify whether certain agricultural production techniques are convenient for a given society, we should first analyse how the organization and technological development of society influence the way in which food is produced in its agricultural sector. Two main characteristics of society that I will examine in this paper are:

1. Population density, or demographic pressure. One of the most obvious aspects of demographic pressure that directly influences the agricultural sector is the amount of arable land per capita available for agricultural production.

2. The level of technological development of a society, or socioeconomic pressure. This can be conveniently quantified by the following two indicators:

   — The average per capita income or average per capita commercial energy consumption ill society. For instance, average per capita commercial energy consumption ranges from less than 1 GJ per year in developing countries to more than 300 GJ per year in the USA.

   — The fraction of total human time available that is allocated to work in the primary sectors of the economy. This fraction ranges from 0,1 ill developing countries to 0.04 ill developed countries. The lower fraction of human time allocated to the primary economic sectors in developed countries is related to the longer education period, the progressive graying of the population, the lighter workload of the labour force, and the expansion of the service sector (up to 60% of the labour force). Of the primary economic sectors, agriculture is one of the most affected in terms of contraction of human time allocation. A cross-section analysis over a sample of 60 countries for the years 1990-1991 shows that the fraction of the labour force engaged in agriculture is inversely correlated with the GDP per capita. Countries with a GDP per capita of more than US$10000 per year have less than 7% of the working force in agriculture, and the richest countries even as little as 2%.

Do population growth (increase in demographic pressure) and changes in energy consumption and time allocation pattern in the economic process (increase in socioeconomic pressure) have implications for the acceptability of techniques of food production in a developed society? In the affirmative, are there numerical indicators that can express the demographic and socioeconomic pressure that society exerts on the production techniques adopted at the farm level?

The demographic pressure or population density of society is simply defined as the ratio. It is well known that population density has been an important factor in shaping the development of agricultural production techniques. Apart from scientific analyses, common sense also suggests that a high demographic pressure in society will favour farming techniques that yield a high food production per unit area.

The effect of demographic pressure on agricultural production can be expressed by the indicator AP_{DP} (= agricultural productivity imposed by demographic pressure), which is defined as the productivity of land (yield of food energy per hectare) 2 that would be needed to obtain a situation of complete food self-sufficiency in society (matching the aggregate food requirement of society). Thus defined numerical assessments of AP_{DP} require knowledge of the aggregate requirement of food in society (considering the food system under analysis as closed), which is in turn determined by the population size of society, the food consumption pattern, and post-harvest losses; The amount of land available for food production, which depends on availability of arable land, characteristics of arable land, and alternative land-uses (dependent on population size and level of technical development).

The effect of socioeconomic pressure on agriculture can be expressed by the indicator AP_{SEP} agricultural productivity imposed by socioeconomic pressure), which is defined as the productivity of labour in agriculture (yield of food energy per hour of labour in the agricultural sector) that would be needed to obtain food self-sufficiency for society.

Considering society's food system as closed, AP_{SEP} is obtained by dividing the yearly aggregate requirement of food in society by the labour time available in a year for generating food supply. The labour time available in a year for generating the food supply depends on the size of the labour force, the unemployment rate, tile fraction of the labour force absorbed by the non-agricultural sectors, and the average work load. The US economy, with only 2% of its workers engaged in agriculture, has an APSEP of 270 MJ of food energy per hour of agricultural labour. This value is well out of the range of possible productivity of fanners in developing countries (less than 10 W 1 and is not even reached by farmers in the European Union. The requirement for food in society is proportional to population size, and therefore to the total human time available to society, whereas internal food production is proportional to the labour time available in the agricultural sector. Hence, in order to achieve and maintain food self-sufficiency, any reduction in the fraction of total labour time allocated to the sector must be compensated by an increase in the amount of food produced per hour of labour in the agricultural sector. According to this rationale, a socioeconomic structure that allocates only a small fraction of its labour force to the agricultural sector will select farming techniques that result in a high food production per unit of labour in the agricultural sector.

The indicators AP_{DP} and AP_{SEP} represent the minimum density of biophysical throughput, per hectare and per hour, respectively, that would be needed on average in the agricultural sector to achieve a state of food self-sufficiency in a defined society. Hence, they are based on characteristics of the socioeconomic system as a whole, and their values can be calculated without any knowledge of what is going on at the level of the fanning system in the society. How, then, can these indicators be related to the actual technological changes occurring at the farm level? Are the values of the indicators AP_{DP} and AP_{SEP}, that characterize the pressure that society exerts on its fanning systems, effectively related to the techniques of food production realized at the farm level? To answer these questions it is useful to compare, for several types of societies, the values of AP_{DP} and AP_{SEP} with actual values achieved in agricultural production in these societies. For instance, the minimum productivity of land imposed by demographic characteristics (AP_{DP}) can he compared with the actual agricultural productivity per hectare (AP_{ha}) for the country as a whole (average), or at the regional level, or for individual farms. The actual productivity of land or yield of food energy per hectare achieved by a country, or farming system is obtained by dividing the aggregate production of food within the defined territory by the amount of land used in the production process, Similarly, expected minimum value of labour productivity in agriculture according to socioeconomic characteristics can be compared with the actual agricultural productivity (throughput of food energy) achieved per hour of labour in agriculture for the country as a whole (average) or for selected farming systems. Such a comparative analysis, at the national level has been carried out for a large sample of countries, to see whether the demographic and socioeconomic pressure are indeed related to the way in which food is actually produced in different socioeconomic contexts. The countries included in the sample are listed in Table 1 and cover a wide range of demographic pressure, socioeconomic pressure, and geographic location.

The analysis shows that even if the two indicators do not indicate mandatory threshold values, they should in any case be considered attractor values, toward which the agricultural system feuds to move in practice (Figs. 2 and 3) 3, If a country has a high demographic pressure it is also likely to have a large production of food energy per hectare in the agricultural sector.

In other words, current technological performance in agriculture in terms of yield per hectare is affected by the existing demographic pressure. Similarly a country experiences considerable socioeconomic pressure it is also likely to have a large production of food energy per hour of work delivered in the agricultural sector. Hence, technological performance in agriculture in terms of labour productivity is affected by the existing socioeconomic pressure. Note that in industrialized societies trade plays a significant role in stabilizing an equilibrium between the requirement and supply of food. Whenever AP_{DP} exceeds AP_{ha} at country level and/or AP_{SEP} exceeds AP_{ha} at country level, society somehow relies on import to offset existing differences between internal food requirements and the internal supply of food realized (both in terms of land and agricultural labour). Obviously, countries with net food export have a value of AP_{DP} that is smaller than their AP_{ha} arid/or a value of AP_{SEP} that is smaller than their AP_{hour}. If the analysis is restricted to countries that had a GNP per capita of less than US$1000 per year in 1991, it becomes clear that in poor countries, where the food system is less open and hence risk aversion is of greater importance, food self-sufficiency in terms of both land and labour productivity in the agricultural sector is more strictly adhered to in practice than in industrialized countries where trade plays a major role.

Countries not achieving self-sufficiency must rely on imports or food aid to guarantee food security for their population. The analysis described above indicates the existence of a link between the variables AP_{DP} and AP_{SEP} on the one hand, that are defined at the hierarchical level of society as a whole, and the variables AP_{DP} and AP_{hour} on the other hand, that are defined and measured at the hierarchical level of farming systems. This link suggests that it is possible to describe technical changes in food production on both levels on the same geometric plane. This plane is defined by two axes, the density of agricultural throughput per hour of labour and the density of agricultural throughput per hectare of land.

Considering the process of agricultural production, changes in AP_{ha} and AP_{hour} reflect changes in technical coefficients of technologies adopted at the farming system level. On the higher level, society as a whole, changes in AP_{DP} and AP_{hour} reflect alterations in the characteristics of the socioeconomic system to which the agricultural sector belongs. Being part of both levels, agriculture must operate in such a way that the combination of values of AP_{DP} and AP_{hour} is compatible with the constraints imposed by society, that is the combination of the values of AP_{DP} and AP_{SEP}. Clearly, farming systems of subsistence societies occupy tin area of the plane that is quite distant from that occupied by farming systems of industrialized countries. In general, farming systems tend to have an actual agricultural performance that is closely related to the set of socioeconomic constraints corresponding to the socioeconomic context in which they are operating. Describing changes in agricultural production on the plane formed by the agricultural throughput per hour of labour and the throughput per hectare of land makes it possible to define trajectories of changes for both socioeconomic variables and technological coefficients in agriculture, which are expected to accompany trends in technological development. For instance, current trends in economic development suggest an economic growth of developing countries that is characterized by increases in both population size and per capita consumption. This will be reflected in a parallel increase in APDP and APSEP. Sooner or later, these changes in demographic and socioeconomic pressure (APDP and APSEP) will be reflected into altered technical coefficients of farming techniques. That is, technological development in agriculture in this situation tends to evolve toward higher values of AP_{DP} and AP_{hour}.

General trends in the evolution of food production techniques for different types of socioeconomic systems can be represented on the two-dimensional plane. Four main types of socioeconomic systems, having different combinations of demographic and socioeconomic pressure:

— *Socioeconomic systems with low demographic and low socioeconomic pressure.* This situation is characterized by an arable land per capita of more than 0.5 ha and a GNP per capita of less than US$10 000 per year. This type of system would include several African countries, such as Burundi.

— Socioeconomic systems, such as the USA, Canada and Australia, with low demographic and high socioeconomic pressure, that is arable land per capita more than 0,5 ha, and a GNP of more than US$10000 per capita per year.

— Socioeconomic systems, such as PR China, and Egypt, with high demographic and low socioeconomic pressure, that is arable land per capita less than 0. 1 ha and GNP per capita less than US$1000 per year. Socioeconomic systems, such as the West European countries, with high demographic and high socioeconomic pressure, that is arable land per capita less than 0.2 ha, and GNP per capita more than US$10000 USD per year.

According to existing trends in population growth and economic development for these four different types of socioeconomic systems, we may expect the following movements in the plane:

— *Societies with low demographic and socioeconomic pressure.* The population is growing faster than the GNP per capita, which means that AP_{DP} will grow faster than APSEP. Hence, they will move toward a situation typical of PR China.

— *Societies with low demographic and high socioeconomic pressure.* Economic development is expected to be maintained and population growth will he relatively slow but steady. On the plane, this means a slow movement toward higher values of AP_{DP}.

— *Societies with high demographic and low socioeconomic pressure.* They tend to show economic growth and maintain if not expand their already huge population size. At national level, an increasing GNP per capita will result in an accelerated absorption of the labour force currently engaged in agriculture by other sectors of the economy. This will inevitably require a major increase in agricultural labour productivity to maintain security. Hence, a movement toward the West European conditions of agricultural production is to be expected.

— *Societies with high demographic and socioeconomic pressure.* These societies have no alternative but to try to maintain a high material standard of living and keep population growth to a minimum. This means a more or less stable and high level of AP_{SEp} and a very slowly increasing value of AP_{DP}.

For these societies, trying to re duce the environmental impact of their food production becomes a major factor. Note that food import from the international market, a must for countries where $AP_{DP} > AP_{ha}$ and/or $AP_{SEP} > AP_{hour}$ is based on the existence of surpluses produced by countries where the relation between these parameters is inverse. At present, the USA, Canada, Australia and Argentina combined 80% of the net export of cereals on the world market.

However, at their present rate of population growth and because of an increasing concern for the environment this surplus may be seriously eroded ill the near future. For instance, the USA is expected to double its population in 60 years. Food import problems will be accelerated if a general concern for the environment will lead to a less intensive agriculture all over the world. At world level, demographic and socioeconomic pressure are certainly expected to increase. It is often overlooked that, at world level, there is no option to import food from elsewhere. When increases in demographic and socioeconomic pressure are not matched by an adequate increase in productivity of land and labour, food imports of the rich will be based on starvation of the poor. Worldwide, arable land per capita is already less than 0.27 ha per capita at present, and is expected to continue to shrink because of population growth.

In addition, arable land is being lost. For instance, during the past 40 years nearly one third of the world's cropland has been abandoned because of soil erosion and other types of degradation. Most of the added land that replaces this loss has come from marginal land made available by deforestation.

High productivity per hectare oil marginal lands requires large amounts of fossil energy-based inputs. This is one of the driving forces increasing the dependence of the world food supply on fossil energy. Many densely-populated countries are not able to meet their internal demand for food without heavy reliance oil fossil energy stocks. "Half of all peasants in Southern China are alive because of the urea cast ladled onto tiny fields—and very few of their children could be born and survive without spreading more of it in the years and decades ahead".

The increasing dependence of the food security of poor densely-populated countries on oil may become problematic in the near future. In the last few years.

PR China and other Asian countries have been among the countries with the fastest growing economics in terms of GNP. At world level, fuelling the economic development of densely-populated countries represents a major drain on natural resources, notably fresh water and fossil energy stocks. Not only do we risk that a dramatic surge in oil demand will augment oil prices, but we may soon face the bottom of the oil barrels.

The goal of achieving sustainability of food production in the future is closely related to the challenge of mediating between the socioeconomic constraints 'imposed' by society, that push for more concentrated and faster agricultural throughputs, and ecological constraints 'imposed' by natural processes that push in the opposite direction.

Natural processes provide the services on which all agricultural production relies, such as water and nutrient supply, soil fertility, pollination, stabilization of atmospheric composition, and regulation of biogeochemical cycles. Respecting ecological constraints means keeping the flows of biomass produced in the agroecosystem by human management within a range of values that is compatible with regulation mechanisms typical of natural ecosystems.

The mutual compatibility between socioeconomic and ecological constraints in agriculture is simply checked by comparing the minimum density of the throughput in the agroecosystem that would be needed to achieve, in an economically viable way, food security with the maximum densities that would still be compatible with the original ecosystem characteristics (keeping the environmental loading below critical thresholds).

A large discrepancy between the actual density of agricultural throughput and the densities imposed by either socioeconomic or ecological constraints indicates unsustainability of agricultural production. Agriculture is ecologically unsustainable when the actual density of agricultural throughput exceeds the critical environmental loading, and socioeconomically unsustainable when the actual density of agricultural throughput is too low to provide food security and/or economic viability for farmers.

It is very possible that the actual agricultural throughput in many developing countries is both ecologically and socioeconomically unsustainable at the same time. The interface of agriculture with the larger socioeconomic system has been discussed in detail in the previous section. In this section, I explore the interface of agriculture with the larger ecological system.

Agriculture can be defined as a human activity that exploits natural processes and natural resources in order to obtain food and other products considered useful by society. The verb 'exploits' suggests that we deal with an alteration. Indeed, within a defined area humans alter the natural distribution of animal and plant populations in order to selectively increase the density of certain flows of biomass that they consider more useful for the socioeconomic system.

Adopting this definition, we implicitly assume that agriculture has to do with manipulation of biodiversity. Following the schema of ecosystem structure proposed by Odum, a natural ecosystem can be seen as a network of matter and energy flows in which nutrients are mainly recycled within the system and solar energy is used to sustain this matter cycling.

The amount of solar energy used for self-organization by the ecosystem is proportional to the speed of its matter cycles (in terrestrial ecosystems this also has to do with availability and circulation of water). Agriculture then involves a boosting of only those matter and energy flows in the network that humans consider beneficial, and an elimination or reduction of the flows that they consider detrimental to their purposes. In (Icing so, humans amplify some of the existing genetic information (that is contained in populations representing 'useful' biomass), and reduce or eliminate some other genetic information. In other words, changes in the hierarchical structure of matter and energy flows in the ecosystem must be coupled to changes in the hierarchical structure of genetic information clustered in species and distributed over taxa in the geographical space altered for agricultural purposes. The hierarchical structure of genetic information can be stabilized only by a quasi-steady-state in nutrient flows, which in turn depends on the ability to use available flows of free energy.

The resonance between coded information in DNA and the metabolism of living systems implies that the reverse must also he true. The particular hierarchical structure of matter and energy flows is determined by the rate of input/output of individual components of the network and their spatial and quantitative distribution; that is, on a large scale, it depends on the hierarchical organization of genetic information. Profound changes in the densities of harvested biomass can have dramatic consequences for the ecosystem structure. The elimination of ecosystem diversity can affects the stability of the biosphere when the scale of human alteration is increased to reach global relevance.

For example, in order to significantly increase yields of biomass (the amount taken away from the agroecosystem per unit of area and time), it is necessary to fight competing species (pests) and to bypass physiological rates of nutrient cycling in the ecosystem (fertilization) when these become limiting. This calls for selection of crop species and varieties that perform well tinder human-managed conditions.

The interaction of the agricultural sector with both ecosystem and society can be studied by following inputs and outputs of the four main steps of the food system cycle:

— Food production, where food is defined as energy and matter compatible with human metabolism.

— Making the food accessible to consumers, where accessible food is defined as meals ready to be consumed according to defined food consumption patterns.

— Consuming food and generating wastes, where wastes are defined as energy and matter no longer compatible with human needs.

— Recycling wastes to agricultural inputs, where agricultural inputs are defined as energy and matter compatible with the agricultural process.

All steps are interconnected, and hence the definition of a flow as an input, an available resource, an accessible resource, or waste, is arbitrary. First of all, the definition depends on the point of view from which the system is analysed. For example, the introduction of trees leads to increased evapotranspiration, which may be negative in terms of less accessible water in the soil, but positive in terms of more available water in the form of rain clouds. Second, the definition depends on the compatibility of the throughput density with the processes regulating the particular step in question.

For example, human excrement of a small Chinese village are a valuable input for agriculture, but the sewage of a big city is a major pollution problem. Whether or not the speed of a throughput at any particular step is compatible with the system as a whole depends on the internal organization of the system. Nobody expects science to make it possible to feed a person 30000 kcal of food per day, about 10 times the normal amount, while keeping that person alive for a long period of time. Why then do many believe it to be possible to increase the productivity of agroecosystems by several times without generating any, negative side-effects on agroecosystem health?

Indeed, the increasing intensity of agricultural throughputs over increasing areas of our planet should be considered a major threat to the conservation of biodiversity. The more the amount of agricultural biomass harvested from a defined population (considering a single species at the time) in an altered ecosystem differs in density from the biomass turnover of a similar species in naturally occurring ecosystems, the larger can be expected to be the environmental impact on the agroecosystem and the energetic cost of the agricultural production.

Indeed, in order to maintain in 'artificially' high density of energy and matter flow in the selected population of a certain species, humans must 'interfere' in the agroecosystem to amplify the genetic information of the selected populations and Suppress genetic information of competing populations. The first action must be backed up by a supply of the required inputs (external fertilizers and irrigation). The second requires active intervention in the life cycles of competing populations, such as slash and bum, tilling of soil, weeding, and pesticide applications.

Extremely high densities of agricultural throughput (per hectare and/or per hour) necessarily require production techniques that ignore the functional mechanism of natural ecosystems, that is the nutrient cycles powered by a linear flow of solar energy. Clearly, individual species, taken one at the time, cannot generate cycles in matter, since they perform only a defined 'job' (occupy a certain niche) in the ecosystem.

Technical progress sooner or later implies a switch from low-input agriculture based on nutrient cycling within the agroecosystem to high-input agriculture in which the throughput density of harvested biomass is directly controlled and maintained at elevated levels through reliance on external inputs that provide linear flows of both nutrients and energy. In low-input agriculture, where the harvested flow of biomass remains within the range of natural turnover values, the activity of production interferes only to a limited extent with the system of controls regulating matter and energy flows in the ecosystem. This form of agriculture requires that at least several distinct species are used in the process of production so as to maintain the internal cycling of natural inputs as a pillar of the agricultural production process.

For humans this implies a poor control over the flow of produced biomass in the agroecosystem, because of the low productivity per hectare when assessing the yield of a particular crop at the time and the variability in productivity in case of perturbations. Especially serious is the problem of facing dramatic increases of demographic pressure. This type of farming cannot work properly with levels of throughput per hectare too high.

On the positive side, direct and indirect biophysical costs of production are low. Few technical inputs are required per unit of output produced and, when population pressure is not too high, the environmental impact of this form of agricultural production is modest. The contrary is true for high-input agriculture. In this case, the harvested flow of biomass is well out of the range of values that is compatible with regulation processes typical of natural ecosystems (when considering yields of a single cultivated species at the time). In high-input agriculture, human management is based on an eradication of the natural structure of controls in the ecosystem.

In fact, where several tonnes of grain have to be produced per hectare and hundreds of kilograms of grain per hour of agricultural labour, natural rates of nutrient cycling and a natural structure of biological communities are unacceptable. In high-input agriculture, not even the, genetic material used for agricultural production is related to the original ecosystem in which production takes place.

In this way, humans keep the flow of produced biomass tightly under control and adjust yields to match increasing demographic and socioeconomic pressure (e.g. the green revolution). However, this control is paid for with a high energetic cost of food production, because humans must continuously provide artificial regulation in the agroecosystem in the form of inputs and defense against undesired, competitive species.

The environmental impact of high-input food production is necessarily large, and involves a dramatic reduction of biodiversity in the altered area, destruction of the entire set of mechanisms that regulated ecosystem functioning before alteration, and several negative side-effects such as on-site and off-site pesticide and fertilizer pollution, soil erosion, and salinization.

How do changes in technical coefficients of farming techniques, aimed at an increase of both land and labour productivity, affect biodiversity use in agriculture.' What is the nature of trade-off between biodiversity use M agriculture on the one hand and productivity of land and/or labour on the other hand? At the level of the farming system the parameters land productivity and labour productivity are interdependent. The existence of this link is illustrated by the possibility to express the parameter labour productivity as, the product of two terms, one of which is the productivity of the land itself:

$$\text{labour productivity (kg hr}^{-1}) = \text{land productivity (kg ha}^{-1}) \times \text{land managed per hour (ha hr}^{-1})$$

Hence labour productivity can be augmented by either increasing the land area managed per hour of labour, increasing the land productivity or a combination of these two solutions. Usually, technical progress induces a combination of the two solutions which are mutually reinforcing.

The land managed per hour of labour can he augmented, while maintaining the same yield per hectare, by a substitution of machine power for human power. This solution pushes for monocultures, because a monocultures because a monoculture facilitates mechanization by, synchronizing and concentrating in time and space the requirement of inputs and the harvest of outputs. With this solution, labour productivity can be raised by using more land per farmer, while avoiding all increase in energy costs per hectare in the form of land-saving inputs such as fertilizers and irrigation. This solution is typical of farming systems in Canada and Australia, where the demographic pressure is low.

Since capital substitution for labour (human power) must have a positive economic return to pay back its own investment, economic feasibility is a crucial factor tor this solution. There are two obvious ways to increase farm revenues in this situation. First, by trying to get the most out of the local conditions by focusing on the crop(s) that show(s) the best comparative advantage in the area.

Second, by trying to get the most out of the capital investment by expanding the area of the farm, carefully concentrating cultivation in favourable spots and avoiding marginal lands. To facilitate mechanization, fanners generally tend to increase the dimension of their fields, by eliminating patchworks and edges, despite the fact that those landscape patterns and elements are known to be important in the maintenance of beneficial insect species for pest control and conservation of biodiversity in general.

Thus, the use of machine power to increase labour productivity will drive technological development in agriculture further toward the exploitation of comparative advantages and economies of scale (e.g. enlargement of farms). The building of specialized technical infrastructures in the area will reinforce the trend of product specialization and generate a mode-locking effect.

While increasing yields per hectare without enlarging the land area managed per unit of labour can be a (temporary) solution to match an increasing demographic pressure, it has limited power to meet the demand for a higher labour productivity. The problems with this solution are evident in PR China, where massive fertilization and irrigation are practiced to augment yields coupled to labour-intensive farming techniques. Although in this way agriculture will feed a densely-populated country, it will provide little support for its socioeconomic development.

The equation of labour productivity indicates that both an increase in land productivity and an increase in land managed per hour of labour can have a positive marginal return in terms of an increased labour productivity and economic return. Therefore, the particular choice of technical inputs in agricultural production—machinery to boost the land managed per unit of labour or fertilizers and irrigation to boost land productivity—will depend on the expected marginal return of capital investments in these two factors. Sparsely-populated developed countries, such as Canada, the USA and Australia, tend to invest more in machinery, even though a certain investment in land-saving inputs is made when the marginal return is high.

Densely-populated developing countries, such as PR China, Egypt and Bangladesh, tend to predominantly invest in fertilizers and irrigation, and invest only little in machinery. Densely-populated wealthy countries, such as Japan, Italy, and the Netherlands, are in a peculiar situation. In these countries, farming can become a 'mission impossible', because of the simultaneous high demographic (arable land per farmer cannot be significantly expanded) and socioeconomic pressure (the need for high labour productivity and economic return in the agricultural sector).

Despite heavy investments in both machinery and land-saving inputs, farmers in these countries still face the unpleasant reality of a land and labour productivity that do not match the minimum outputs imposed by their demographic and socioeconomic pressure.

For instance, it is impossible for a Japanese farmer to produce hundreds of kilograms of grain per hour of labour by working 1800 hours per year on just one hectare of arable land. When biophysical constraints of this type exist there are no other Ways out than either resorting to a system of government subsidies (transferring money from other economic sectors to agriculture) to guarantee acceptable farmers' income (comparable to that obtained in other economic sectors) and/or importing food to guarantee food security to the country.

The need for high labour productivity (socio-economic pressure) and eventually land productivity (demographic pressure) at the level of society, reflected into changes in technical coefficients at the farm level, is intricately linked to changes occurring in the energy budget of socioeconomic systems to which the agricultural system belongs. Economic development, assessed at the socioeconomic level, inevitably results in a continuously increasing opportunity cost of human labour and capital at the farm level, and pushes technical development of the farming system towards mechanization, specialization, economies of scale, linearization of nutrient inputs in the agroecosystem, and dependence on price support and other subsidies.

This chain of changes, well-known in the field of agricultural economics though generally described from a different perspective, is the main cause of the abandonment of traditional farming strategies, and the consequent loss of biodiversity in agriculture. Most notably, two important trends follow from this process of technological change: adoption of monocultures, and reduced use of by-products, such as crop residues and manure, in the agricultural process.

Natural ecosystems (e.g. mature natural forest) are able to sustain a large amount of biomass, but with a net primary productivity that is relatively low compared to agroecosystems. Traditional farming systems based on gardening in natural forests can produce large quantities of biomass per hectare on the cleared areas, but only with a long cycle of rotation. When the entire area involved iii the rotation is considered, overall crop productivity per hectare is quite low.

Other systems of traditional farming based on inter-cropping, multiple cropping and animal husbandry have proved to be highly productive in terms of food supply per hectare whenever ample fresh water is available. How ever, the combined exploitation of diverse natural resources generally requires labour-intensive manage merit. Rural communities in China and elsewhere in East Asia, masters in the integrated use of natural resources, have been able to feed themselves and pay taxes in form of part of the harvest in spite of a limited endowment of land, for centuries.

However, the particular techniques of multiple cropping, animal husbandry and integrated use of natural re sources implied that more than 80% of the population was engaged in agriculture. Moreover, note that the population of China did not surpass 600 million until synthetic fertilizers were made available to China m the late 1970s. Whenever socioeconomic development calls for a reduction of the labour force in agriculture, the society must confront the need for an increase in farm labour productivity.

For example, the yearly amount of food consumed by a US citizen is produced with less than 20 h of farm labour. Up to the present, this level of agricultural labour productivity has only been achieved by making use of monoculture. Mounting demographic and socioeconomic pressure can make traditional techniques, such as combined production of cereals and legumes, and rotation with green manure, no longer feasible unless *ad hoc* policies are established to provide government support

Recycling of natural inputs within the farming system is abandoned as soon as this becomes too costly in terms of land and/or human labour requirement, provided of course that alternative technical inputs for use in agriculture are available on the market. For example, the trend of abandoning cultivation of green manure in favour of food grain when artificial fertilizer is accessible is discussed in detail by Smil. In general, the decision whether or not to abandon the recycling of natural inputs will depend on the amount of energy or nutrients gained by recycling, divided by the extra hours of labour required, the opportunity-cost of labour, and the price of technical inputs used as alternatives.

When the inputs made available per hour of labour by recycling are significantly less than the inputs that could be bought on the market with the economic return of one hour of labour allocated to off-farm activities, then recycling will generally be abandoned. In this case, linearization of nutrient inputs is more convenient in terms of economic cost/benefit. Within a market economy, the choice to specialize and rely on monocultures often leaves no alternative but to also switch to linearization of nutrient inputs in the form of fertilization.

The continuous and fast changes in the socioeconomic context of farming are driving technological progress in agriculture toward solutions (densities of agricultural throughputs) that are increasingly incompatible with ecological processes. The need to keep pace with mounting demographic and socioeconomic pressure is affecting agricultural development all over the world, and threatens biodiversity in both wild ecosystems (their area is shrinking) and managed ecosystems (the intensity of alteration is increasing).

Criteria adopted for choosing and developing farming techniques have generally been based on socioeconomic constraints only (costs/benefits referring to economic growth) and therefore have been biased in favour of continuously increasing densities of harvested biomass.

Ecological constraints imposed by the characteristics of natural ecosystems, that would put an upper limit to the densities of harvested biomass, tend to be ignored. This is for the simple reason that socioeconomic constraints to agriculture leading to technological 'progress' have universal application (technical coefficients refer to unit operations in artificial environments and can easily he generalized in terms of monetary costs/ benefits), whereas ecological constraints are location-specific and extremely difficult to define and quantify. The consequent imbalance between the short-term and long-term (ecological) perspective when deciding how to manage agricultural production may have disastrous consequences given the immense scale of agricultural activities worldwide approximately 50% of the terrestrial area is devoted to agriculture, while in temperate ecosystems agriculture occupies 70% of the land.

As Hollings has pointed out, in biological systems a general homogenization of activities over a large scale is a disaster waiting to happen: it dramatically increases the risk that a perturbation damaging to that pattern will occur and be amplified over the whole system.

In conclusion, strict respect the pattern of natural process in the agroecosystem to preserve biodiversity would slow down the throughput of food production per hectare of land and per hour of labour. When these throughputs become so low so as to become incompatible with socioeconomic characteristics this solution becomes economically non-viable.

On the other hand, matching demographic and socioeconomic pressure above certain ecological thresholds leads to ecological unsustainability. For example, the production of tonnes of grain per hectare per year and hundreds of kilograms of grain per hour of labour with less than 0.2 hectare of arable land per capita can only be obtained in an ecologically unsustainable way through linearization of nutrient flows in the agroecosystem, that is continuous harvests (no rotation) and heavily reliance on technical inputs such as machinery, fertilizer, pesticides, and irrigation.

Discussions of sustainability of world agricultural production should carefully consider both ecological constraints such as the possible limiting role of land, water, and other natural resources, and socioeconomic constraints to food production.

Current strategies of technological development in agriculture not only appear unable to solve problems of food security in the long term, but also have been increasingly malfunctioning in the last decade, both in developed and in many developing countries.

In developed countries this is because excessive socioeconomic pressure has caused farms and fanners to disappear, while remaining farming systems employ production techniques that are detrimental to ecological equilibria, and in developing countries it is because excessive demographic pressure has caused poverty in rural areas to increase and food production per capita to decrease.

Paradoxically, this evident malfunctioning is perhaps also a reason for optimism. After all, when dealing with human systems, cultural mode-locking is the most persistent and important cause of lack of adaptability in response to changing boundary conditions.

As concerns agriculture, however, there is an increasing, though not yet general, consensus among farmers, scientists, politicians and the general public that we somehow have to change the way we look at and practice agriculture. In terms of opportunities for future changes in farming strategies, the situation is dramatically different in developed and developing countries.

In developed countries, where high-input agriculture is the rule rather than the exception, there is some room for maneuvering to get out of the current path of development. Intensive agriculture has been driven by a strong and fast socioeconomic development and based on the creation of a 'risk-free environment' for farmers and investors in the agro-industry.

This risk-free environment, that was and still is financially supported by the rest of society, has been realized through guaranteed prices, input subsidies, disaster, relief, etc. Indeed, modem high-input agriculture depends economically on the rest of society, that mobilizes economic resources to maintain its agricultural sector. We may thus conclude that in developed countries the economic resources are available to support the agricultural sector, but that they should be spent in a different way.

Developing countries have never disposed over sufficient economic resources to generate a risk-free environment for farmers and investors in high-input agriculture. On the contrary, the agricultural sector is usually taxed to support other sectors of the economy.

In the absence of government support, risks implied by excessive simplification of ecosystems typical of high-input agriculture are transferred to individual farmer families. This makes such a solution less attractive, especially for poor farmers operating in marginal areas. Nevertheless, the continuous increase in demographic and/or socioeconomic pressure as well as access to Western technologies, has pushed for the adoption of high-input techniques in local attempts to augment agricultural throughputs. Given the rapidly growing population in developing countries (actual increases in demographic pressure) and the gap still to fill before reaching desired income levels (potential socioeconomic pressure), we can expect the density of agricultural throughputs to increase in most, if not all, developing countries, even though this could be destructive for the environment in many cases.

In spite of this wide difference, there is a striking similarity between high-input and low-input agriculture. In both cases, food security totally depends on the availability of fresh water, fertile soil, pollination, natural recycling of nutrients and pollutants, biological pest control, and nitrogen fixation. These are services that cannot be substituted by injection of technical capital, and that are heavily dependent on biodiversity.

The neoclassic economic paradigm endorses the idea that unlimited trade is a useful tool to enhance world economic growth. Adopting this tool, individual countries can overcome biophysical constraints in the form of local shortages of natural resources that limit the expansion of their economics. According to this view, trade enables a better use of natural resources through complementation of economies (comparative advantages).

However, this view implicitly implies the adoption of the entire world as the hierarchical level at which the economic process is assessed and optimized. This level is not necessarily the best possible choice, especially when dealing with agricultural production.

In fact, the use of a smaller space-time scale for defining strategies of technological development also has positive effects. First, a smaller scale makes it easier to involve local people in the decision-making process and increase the responsibility of local communities for resource management.

Thus, decisions related to sustainability should be made as close as possible to local people, and with the participation of all major stake-holders. Apart from the ethical aspect, this would make it possible to benefit from indigenous knowledge of the nature of the problem of sustainability dealt with. Second, internalizing most of the external services on which economic systems rely (by reducing trade) makes it easier for the system to respond quickly to changes and signals coming from surrounding ecosystems, in spite of their intricate and varying nature. Societies heavily relying on trade often cannot see the importance of the ecological services on which they directly and indirectly rely. Therefore, energy expenditure in transportation from distant ecosystems should be justified only in cases of substantial comparative advantages. Third, the goal of harmonizing energy and material circulation with the characteristics of local ecosystems (decreasing the environmental impact of economic activities) can he more easily achieved it' the space-time closure of the activities of production and consumption in the society is similar to the space-time closure of the matter cycles occurring in local ecosystems. These points suggest that the enlargement of scale of economic systems obtained by heavy trading can also induce negative effects, both on the stability of the biosphere and on the integrity of local socioeconomic systems. There is therefore also a rationale for policies aiming at increasing the level of self-sufficiency and addressing the issue of social equity at the level of national and regional economic systems.

When defining the terms of reference for sustainable agriculture, biodiversity should be considered as an output rather than as an input in agriculture. Agriculture should not only produce food, but also preserve a healthy environment. The latter task, grossly neglected, involves guaranteeing a higher level of cycling of nutrients, better control of pollution of fresh water supplies, and a sound management of the landscape in general. For all these tasks preservation of biodiversity is essential.

Conservation of biodiversity, however, only makes sense within a large-scale planning, both in terms of time and space. For instance, if socioeconomic pressure imposes high-input agriculture, then we should exploit only those terrestrial ecosystems that are better adapted to being altered, including rotation and fallow. Improving long-term sustainability of agricultural production implies more emphasis on ecological constraints, even if this implies reducing the short-term efficiency of agricultural production. Indeed, if the 510 year average discount period typically adopted by economic analyses is not abandoned, then there is little chance that assessments of costs and benefits referring to biodiversity will be seriously considered.

Moreover, the perception of the trade-off between long-term ecosystem health and short-term maximization of production is affected by the level of development. In countries with a high material standard of living, citizens often consider it acceptable to reduce economic returns for the benefit of the environment, while developing countries do not have room to reduce economic returns. They have no other option but to boost short-term returns at the cost of long-term sustainability.

At the moment of discussing agricultural policies in view of sustainability, the small space-time scale typical of economic analyses should always be balanced by a long-term and worldwide view. But is this a feasible idea? Developed countries might be prepared to sacrifice part of their own economic return for the benefit of their immediate surroundings, but doing this for the benefit of other, distant societies and ecosystems is something quite different. The fact that all socioeconomic systems, ecosystems and agroecosystems form part of a larger whole, the functioning of which is affected by the health of all its individual components, often appears to be beyond our capacity and willingness of understanding. The 'biodiversity revolution in agriculture' should result in a definition of a set of new optimizing parameters for agricultural techniques.

15

ASEAN Agreement on the Conservation of Nature and Natural Resources

The Government of Negara Brunei Darussalam, The Government of the Republic of Indonesia, The Government of Malaysia, The Government of the Republic of Philippines, The Government of the Republic of the Singapore and The Government of the Kingdom of Thailand, Member States of the Association of South East Asian Nations (ASEAN).

Recognizing the importance of natural resources for present and future generations;

Conscious of their ever-growing value from a scientific, cultural, social and economic point of view;

Conscious also that the interrelationship between conservation and socio-economic development implies both that conservation is necessary to ensure sustainability of development, and that socio-economic development is necessary for the achievement of conservation on a lasting basis;

Recognizing the interdependence of living resources, between them and with other natural resources, within ecosystems of which they are part;

Wishing to undertake individual and joint action for the conservation and management of their living resources and the other natural elements on which they depend;

Recognizing that international co-operation is essential to attain many of these goals;

Convinced that an essential means to achieve such concerted action is the conclusion and implementation of an Agreement;

Have agreed as follows:

Chapter I

Conservation and Development

Article 1

Fundamental Principle

1) The Contracting Parties, within the framework of their respective national laws, undertake to adopt singly, or where necessary and appropriate through concerted action, the measures necessary to maintain essential ecological processes and life-support systems, to preserve genetic diversity, and to ensure the sustainable utilization of harvested natural resources under their jurisdiction in accordance with scientific principles and with a view to attaining the goal of sustainable development.

2) To this end they shall develop national conservation strategies, and shall co-ordinate such strategies within the framework of a conservation strategy for the Region.

Article 2

Development Planning

1) The Contracting Parties shall take all necessary measures, within the framework of their respective national law, to ensure that conservation and management of natural resources are treated as an integral part of development planning at all stages and at all levels.

2) To that effect they shall, in the formulation of all development plans, give as full consideration to ecological factors as to economic and social ones.

3) The Contracting Parties shall, where necessary, take appropriate action with a view to conserving and managing natural resources of significant importance for two or several Contracting Parties.

Chapter II

Conservation of Species and Ecosystems

Article 3

Species - genetic diversity

1) The Contracting Parties shall, wherever possible, maintain maximum genetic diversity by taking action aimed at ensuring the survival and promoting the conservation of all species under their jurisdiction and control.

2) To that end, they shall adopt appropriate measures to conserve animal and plant species whether terrestrial, marine and freshwater, and more specifically

   (a) conserve natural, terrestrial, freshwater and coastal or marine habitats;

(b) ensure sustainable use of harvested species;

(c) protect endangered species;

(d) conserve endemic species; and

(e) take all the measures in their power to prevent the extinction of any species or sub-species.

3) In order to fulfil the aims of the preceding paragraph of this Article the Contracting Parties shall, in particular, endeavour to

(a) create and maintain protected areas;

(b) regulate the taking of species and prohibit unselective taking methods;

(c) regulate and, where necessary, prohibit the introduction of exotic species;

(d) promote and establish gene banks and other documented collections of animal and plant genetic resources.

Article 4

Species - Sustainable use

The Contracting Parties shall pay special attention to harvested species, and, to that effect, shall endeavour to

1) develop, adopt and implement management plans for those species, based on scientific studies and aiming at

(a) preventing decrease in the size of any harvested population to levels below those which ensure its stable recruitment and the stable recruitment of those species which are dependent upon, or related to them;

(b) maintaining the ecological relationship between harvested, dependent and related populations of living resources of the ecosystem considered;

(c) restoring depleted populations to at least the levels referred to in sub-paragraph (1) of this paragraph;

(d) preventing changes or minimizing risk of changes in the ecosystem considered which are not reversible over a reasonable period of time.

2) Take the appropriate and necessary legislative and administrative measures on harvesting activities in the light of their national interest whereby

(a) such activities must conform to the management plans referred to above;

(b) the conduct of such activities is controlled by a permit system;

(c) all indiscriminate means of taking and the use of all means capable of causing local extinction of, or serious disturbance to, populations of a species or related species are prohibited;

(d) such activities are prohibited or strictly regulated at certain periods, seasons or places of importance in the life cycle of the species;

(e) such activities may be regulated more strictly, temporarily or locally in order to assist restoration of population levels or counterbalance any threat caused by special circumstances;

(f) special measures, such as restocking, are provided for whenever the conservation status of a species so warrants;

(g) trade and possession of specimens or products of specimens are regulated whenever such regulations meaningfully contribute to the implementation of the harvesting regulations.

Article 5

Species - endangered and endemic

1) Appendix 1 to this Agreement shall list endangered species recognized by the Contracting Parties as of prime importance to the Region and deserving special attention. The Appendix shall be adopted by a meeting of the Contracting Parties.

Accordingly, Contracting Parties shall, wherever possible.

(a) prohibit the taking of these species, except for exceptional circumstances by special allowance from the designated authorities of the Contracting Parties;

(b) regulate the trade in and possession of specimens and products of those species accordingly;

(c) especially protect habitat of those species by ensuring that sufficient portions are included in protected areas;

(d) take all other necessary measures to improve their conservation status, and restore their populations to the highest possible level.

2) Each Contracting Party shall, wherever possible, apply the above measures to species endangered at national level.

3) The Contracting Parties recognize their special responsibility in respect of species that are endemic to areas under their jurisdiction and shall undertake accordingly to take wherever possible, all the necessary measures to maintain the population of such species at the highest possible level.

Article 6

Vegetation Cover and Forest Resources

1) The Contracting Parties shall, in view of the role of vegetation and forest covers in the functioning of natural ecosystems, take all necessary measures to ensure the conservation of the vegetation cover and in particular of the forest cover on lands under their jurisdiction.

2) They shall, in particular, endeavour to

(a) control clearance of vegetation; endeavour to prevent bush and forest fires; prevent overgrazing by, inter alia, limiting grazing activities to periods and intensities that will not prevent regeneration of the vegetation;

(b) regulate mining and mineral exploration operations with a view to minimizing disturbance of vegetation and to requiring the rehabilitation of vegetation after such operations;

(c) set aside areas as forest reserves, inter alia, with a view to conserve the natural forest genetic resources;

(d) in reforestation and afforestation planning avoid as far as possible monoculture causing ecological imbalance;

(e) designate areas whose primary function shall be the maintenance of soil quality in the catchment considered and the regulation of the quantity and quality of the water delivered from it;

(f) ensure, to the maximum extent possible the conservation of their natural forests, particularly mangroves with a view, inter alia, to maintaining maximum forest species diversity;

(g) develop their forestry management plans on the basis of ecological principles with a view to maintaining potential for optimum sustained yield and avoiding depletion of the resource capital.

Article 7

Soil

1) The Contracting Parties shall, in view of the role of soil in the functioning of natural ecosystems, take measures, wherever possible towards soil conservation, improvement and rehabilitation; they shall in particular, endeavour to take steps to prevent soil erosion and other forms of degradation, and promote measures which safeguard the processes of organic decomposition and thereby its continuing fertility.

2) To that effect, they shall, in particular, endeavour to

(a) establish land use policies aimed at avoiding losses, of vegetation cover, substantial soil losses, and damages to the structure of the soil;

(b) take all necessary measures to control erosion, especially as it may affect coastal or freshwater ecosystems lead to siltation of downstream areas such as lakes or vulnerable ecosystems such as coral reefs, or damage critical habitats, in particular that of endangered or endemic species;

(c) take appropriate measures to rehabilitate eroded or degraded soils including rehabilitation of soil affected by mineral exploitation.

Article 8

Water

1) The Contracting Parties shall, in view of the role of water in the functioning of natural ecosystems, take all appropriate measures towards the conservation of their underground the surface water resources.

2) They shall to that effect, in particular, endeavour to

   (a) undertake and promote the necessary hydrological research especially with a view to ascertaining the characteristics of each watershed;

   (b) regulate and control water utilization with a view of achieving sufficient and continuous supply of water for, inter alia, the maintenance of natural life supporting systems and aquatic fauna and flora;

   (c) when planning and carrying out water resource development projects take fully into account possible effects of such projects on natural processes or on other renewable natural resources and prevent or minimize such effects.

Article 9

Air

The Contracting Parties shall, in view of the role of air in the functioning of natural ecosystems, endeavour to take all appropriate measures towards air quality management compatible with sustainable development.

Chapter III

Conservation of Ecological Processes

Article 10

Environmental Degradation

The Contracting Parties, with a view to maintaining the proper functioning of ecological processes, undertake, wherever possible, to prevent, reduce and control degradation of the natural environment and , to this end, such endeavour to undertake, in addition to specific measures referred to in the following article;

(a) to promote environmentally sound agricultural practice by, inter alia, controlling the application of pesticides, fertilizers and other chemical products for agricultural development schemes, in particular for wetland drainage or forest clearance, pay due regard to the need to protect critical habitats as well as endangered and economically important species;

(b) to promote pollution control and the development of environmentally sound industrial processes and products;

(c) to promote adequate economic or fiscal incentives for the purposes of sub-paragraphs (a) and (b) above;

(d) as far as possible to consider the originator of the activity which may lead to environmental degradation responsible for its prevention, for rehabilitation and remedial measures required;

(e) to take into consideration, when authorizing activities likely to affect the natural environment, the foreseeable interactions between the new activities proposed and those already taking place in the same area, and the result of such interactions on the air, waters and soils of the area;

(f) to pay particular attention to the regulation of activities which may have adverse affects on processes which are ecologically essential or on areas which are particularly important or sensitive form an ecological point of view, such as the breeding and feeding grounds of harvested species.

Article 11

Pollution

The Contracting Parties, recognizing the adverse effect that polluting discharges or emissions may have on natural processes and the functioning of natural ecosystems as well as on each of the individual ecosystem components, especially animal and plants species, shall endeavour to prevent, reduce and control such discharges, emissions or applications, in particular by

(a) submitting activities likely to cause pollution of the air, soil, freshwater, or the marine environment, to controls which shall take into consideration both the cumulative effects of the pollutants concerned and the self-purificating aptitude of the recipient natural environment;

(b) making such controls conditional on, inter alia, appropriate treatment of polluting emissions; and

(c) establishing national environmental quality monitoring programmes, particular attention being paid to the effects of pollution on natural ecosystems, and co-operation in such programmes for the Region as a whole.

Chapter IV
Environmental Planning Measures

Article 12

Land Use Planning

1) The Contracting Parties shall, wherever possible in the implementation of their development planning, give particular attention to the national allocation of land usage. They shall endeavour to take the necessary measures to ensure the integration of natural resources conservation into the land use planning process and shall, in the preparation and implementation of specific land use plans at all levels, give as full consideration as possible to ecological factors as to economic and social ones. In order to achieve optimum sustainable land use, they undertake to base their land use plans as far as possible on the ecological capacity of the land.

2) The Contracting Parties shall, in carrying out the provisions of paragraph (1) above, particularly consider the importance of retaining the natural high productivity of areas such as coastal zones and wetlands.

3) They shall, where appropriate, coordinate their land use planning with a view to conserving and managing natural resources of significant importance for two or several Contracting Parties.

Article 13

Protected Areas

1) The Contracting Parties shall as appropriate establish, in areas under their jurisdiction, terrestrial, freshwater, coastal or marine protection areas for the purpose of safeguarding

- the ecological and biological processes essential to the functioning of the ecosystems of the Region;
- representative samples of all types of ecosystem of the Region;
- satisfactory population levels for the largest possible number of species of fauna and flora belonging to those ecosystems;
- areas of particular importance because of their scientific, educational, aesthetic, or cultural interest;

and taking into account their importance in particular as:

- the natural habitat of species of fauna and flora, particularly rare or endangered or endemic species:
- zones necessary for the maintenance of exploitable stocks of economically important species;
- pools of genetic material and safe refuges for species, especially endangered ones;
- sites of ecological, aesthetic or cultural interest;
- reference sources of scientific research;

They shall, in particular, take all measures possible in their power to preserve those areas which are of an exceptional character and are peculiar to their country or the Region as well as those which constitute the critical habitats of endangered or rare species, of species that are endemic to a small areas and of species that migrate between countries of Contracting Parties.

2) Protected areas established pursuant to this Agreement shall be regulated and managed in such a way as to further the objectives of the purpose of which they have been created.

Contracting Parties shall, wherever possible, prohibit within such protected areas activities which are inconsistent with such objectives.

3) Protected areas shall include

(a) *National Parks*

i) This expression denotes natural areas that are sufficiently large to allow for ecological self-regulation of one or several ecosystems, and which have not been substantially altered by human occupation or explo-itation.

ii) National Parks shall be placed under public control, their boundaries shall not be altered nor shall any portion of any National Park be alienated except by the highest competent authority.

iii) National Parks shall be dedicated to conservation and to scientific, educational and recreational uses and the common welfare of the people.

(b) *Reserves*

(i) This expression denotes areas set aside for the purpose of preserving a specific ecosystem, the critical habitat of certain species of fauna or flora, a water catchment area or for any other specific purpose relating to the conservation of natural resources or objects or areas of scientific, aesthetic, cultural, educational or recreational interest.

ii) After reserves have been established their boundaries shall not be altered nor shall any portion of such reserves be altered nor shall any portion of such reserves be alienated except by the authority establishing them or by higher authority.

iii) Reserves shall be dedicated to the purposes for which they have been created and, in the light of the national interests of the Contracting Parties, any activity inconsistent with such purposes shall be prohibited.

4) Contracting Parties shall, in respect of any protected areas established pursuant to this Agreement

(a) prepare a management plan and manage the area on the basis of this plan;

(b) establish, wherever appropriate, terrestrial or aquatic buffer zones that shall be located around protected areas and which, in the case of marine areas, may include coastal land areas or watersheds of rivers flowing into the protected area; in such buffer zones all activities that may have harmful consequences on the ecosystems that such areas purport to protect shall be prohibited or regulated and activities which are consistent with the purpose of the protected area shall be promoted.

5) Contracting Parties shall, in respect of any protected area established pursuant to this Agreement, endeavour to

(a) prohibit the introduction of exotic animal or plant species;

(b) prohibit the use or release of toxic substances or pollutants which could cause disturbance or damage to protected ecosystems or to the species they contain;

(c) to the maximum extent possible, prohibit or control any activity exercised outside protected areas when such an activity is likely to cause disturbance or damage to the ecosystems or species that such protected areas purport to protect.

6) Contracting Parties shall co-operate in the development of principles, objectives, criteria and guidelines for the selection, establishment and management of protected areas in the Region with a view to establishing a co-ordinated network of protected areas throughout the Region giving particular attention to those of regional importance. An Appendix containing such principles, objectives, criteria and guidelines shall be drawn up in the light of the best scientific evidence as adapted to the conservation requirements of the Region and shall be adopted by a meeting of Contracting Parties.

7) In addition to the establishment of the protected areas referred to in paragraph 3 of this Article, Contracting Parties shall promote, through the adoption of appropriate measures, the conservation of natural areas by private owners, community or local authorities.

Article 14

Impact Assessment

1) The Contracting Parties undertake that proposals for any activity which may significantly affect the natural environment shall as far as possible be subjected to an assessment of their consequences before they are adopted, and they shall take into consideration the results of this assessment in their decision-making process.

2) In those cases where any such activities are undertaken, the Contracting Parties shall plan and carry them out so as to overcome or minimize any assessed adverse effects and shall monitor such effects with a view to taking remedial action as appropriate.

Chapter V

National Supporting Measures

Article 15

Scientific Research

The Contracting Parties shall individually or in co-operation with other Contracting Parties or appropriate international organizations, promote and, whenever possible, support scientific and technical programmes of relevance to the conservation and management of natural resources, including monitoring, research, the exchange of technical information and the evaluation of results.

Article 16

Education, Information and Participation of the Public, Training

1) The Contracting Parties shall endeavour to promote adequate coverage of conservation and management of natural resources in education programmes at all levels.

2) They shall circulate as widely as possible information on the significance of conservation measures and their relationship with sustainable development objectives, and shall, as far as possible, organize participation of the public in the planning and implementation of conservation measures.

3) Contracting Parties shall endeavour to, individually or in co-operation with other Contracting Parties or appropriate international organizations, develop the programmes and facilities necessary to train adequate and sufficient scientific and technical personnel to fulfil the aims of this Agreement.

Article 17.

Administrative Machinery

1) The Contracting Parties shall identify or maintain the administrative machinery necessary to implement the provisions of this Agreement, and, where several governmental institutions are involved, create the necessary co-ordinating mechanism for the authorities dealing with designated aspects of the environment.

2) They shall endeavour to allocate sufficient funds to the task necessary for the implementation of this Agreement, as well as sufficient qualified personnel with adequate enforcement powers.

Chapter VI
International Co-operation

Article 18

Co-operative Activities

1) The Contracting Parties shall co-operate together and with the competent international organizations, with a view to co-ordinating their activities in the field of conservation of nature and management of natural resources and assisting each other in fulfilling their obligations under this Agreement.

2) To that effect, they shall endeavour

   (a) to collaborate in monitoring activities;

   (b) to the greatest extent possible, co-ordinate their research activities;

   (c) to use comparable or standardized research techniques and procedures with a view to obtaining comparable data;

   (d) to exchange appropriate scientific and

technical data, information and experience, on a regular basis;

(e) whenever appropriate, to consult and assist each other with regard to measures for the implementation of this Agreement.

3) In applying the principles of co-operation and co-ordination set for the above, the Contracting Parties shall forward to the Secretariat

(a) information of assistance in the monitoring of the biological status of the natural living resources of the Region;

(b) Information, including reports and publications of a scientific, administrative or legal nature and, in particular, information on

— measures taken by the Parties in pursuance of the provisions of this Agreement;

— the status of species included in Appendix 1;

— any other matter to which the Conference of the Parties may give special priority.

Article 19

Shared Resources

1) Contracting Parties that share natural resources shall co-operate concerning their conservation and harmonious utilization, taking into account the sovereignty, rights and interests of the Contracting Parties concerned in accordance with generally accepted principles of international law.

2) To that end, they shall, in particular

(a) co-operate with a view to controlling, preventing, reducing or eliminating adverse environmental effects which may result in one Contracting Party from the utilization of such resources in another Party;

(b) endeavour to conclude bilateral or multilateral agreements in order to secure specific regulation of their conduct in respect of the resources concerned;

(c) as far as possible make environmental assessments prior to engaging in activities with respect of shared natural resources which may create a risk of significantly affecting the environment of another sharing Contracting Party or other sharing Contracting Parties;

(d) notify in advance the other sharing Contracting Party or the other sharing Contracting Parties of pertinent details of plans to initiate, or make a change in, the conservation or utilization of the resource which can reasonably be expected to affect significantly the environment in the territory of the other Contracting Party or Contracting Parties;

(e) upon request of the other sharing Contracting Party or sharing Contracting Parties, enter into consultation concerning the above mentioned Plans;

(f) inform the other sharing Contracting party or other sharing Contracting Parties of emergency situations or sudden grave natural events which may have repercussions on their environment;

(g) whenever appropriate, engage in joint scientific studies and assessments, with a view to facilitating co-operation with regard to environmental problems related to a shared resource, on the basis of agreed data.

3) Contracting Parties shall especially co-operate together and, where appropriate, shall endeavour to co-operate with other Contracting Parties, with a view to

(a) the conservation and management of

— border or contiguous protected areas;

— shared habitats of species listed in Appendix 1;

— shared habitats of any other species of common concern;

(b) the conservation, management and, where applicable, regulation of the harvesting of species which constitute shared resources;

— by virtue of their migratory character, or

— because they inhabit shared habitats.

Article 20

Transfrontier Environmental Effects

1) Contracting Parties have in accordance with generally accepted principles of international law the responsibility of ensuring that activities under their jurisdiction or control do not cause damage to the environment of the natural resources under the jurisdiction of other Contracting Parties or of areas beyond the limits of national jurisdiction.

2) In order to fulfil this responsibility, Contracting Parties shall avoid to the maximum extent possible and reduce to the minimum extent possible adverse environmental effects of activities under their jurisdiction or control, including effects on natural resources, beyond the limits of their national jurisdiction.

3) To that effect, they shall endeavour

(a) to make environmental impact assessment before engaging in any activity that may create a risk of significantly affecting the environment or the natural resources of another Contracting Party or the environment or natural resources beyond national jurisdiction;

(b) to notify in advance the other Contracting Party or Contracting Parties concerned of pertinent details of plans to initiate, or make a change in, activities which can reasonably be expected to have significant effects beyond the limits of national jurisdiction;

(c) to enter into consultation concerning the above-mentioned plans upon request of the Contracting Party or Contracting Parties in question;

(d) to inform the Contracting Party or Contracting Parties in question of emergency situations or sudden grave natural events which may have repercussion beyond national jurisdiction.

4` contracting Parties shall, in particular, endeavour to refrain from actions which might directly or indirectly adversely affect wildlife habitats situated beyond the limits of national jurisdiction, especially habitats of species listed in Appendix 1 or habitats included in protected areas.

Chapter VII
International Supporting Measures

Article 21

Meeting of the Contracting Parties

1) Ordinary meetings of the Contracting Parties shall be held at least once in three years, in as far as possible in conjunction with appropriate meetings of ASEAN, and extraordinary meetings shall be held at any other time, upon the request of one Contracting Party provided that such request is supported by at least one other Party.

2) It shall be the function of the meetings of the Contracting Parties, in particular

   (a) to keep under review the implementation of this Agreement and the need for other measures, in particular the Appendices;

   (b) to adopt review and amend as required any Appendix to this Agreement;

   (c) to consider reports submitted by the Contracting Parties in accordance with Article 28 or any other information which may be submitted by a Party, directly or through the Secretariat;

   (d) to make recommendations regarding the adoption of any Protocol or any amendment to this Agreement;

   (e) to establish working groups or any other subsidiary body as required to consider any matter related to this Agreement;

   (f) to consider and undertake any additional action, including the adoption of financial rules, that may be required for the achievement of the purposes of this Agreement.

Article 22

Secretariat

On the coming into force of this Agreement the Contracting Parties shall designate the Secretariat responsible for carrying out the following functions:

   (a) to convene and prepare the meetings of Contracting Parties;

   (b) to convene diplomatic conferences for the purpose of adopting Protocols;

   (c) to transmit to Contracting Parties notification, reports and other information received in accordance with this Agreement;

(d) to consider inquiries by, and information from, the Contracting Parties, and to consult with them on questions relating to this Agreement.

(e) to perform such other functions as may be assigned to it by the Contracting Parties;

(f) to ensure the necessary co-ordination with other competent international bodies and in particular to enter into such administrative arrangements as may be required for the effective discharge of the secretariat functions.

Article 23

National Focal Points

In order to facilitate communications with other Parties and the Secretariat, the Contracting Parties shall designate an appropriate national agency or institution responsible for co-ordinating matters arising from consultations and channeling communications between Contracting Parties or with the Secretariat.

Chapter VIII

Final Clauses

Article 24

Adoption of Protocols

1) The Contracting Parties shall co-operate in the formulation and adoption of Protocols to this Agreement, prescribing agreed measures, procedures and standards for the implementation of this Agreement.

2) The Contracting Parties, at a diplomatic conference, may adopt Protocols to this Agreement.

3) The Protocols of this Agreement shall be subject to acceptance and shall enter into force on the thirtieth day after the deposit with the Depositary of the Instruments of Acceptance of all the Contracting Parties.

Article 25

Amendment of the Agreement

1) Any Contracting Party to this Agreement may propose amendments to the Agreement. Amendments shall be adopted by a diplomatic conference which shall be convened at the request of the majority of the Contracting Parties.

2) Amendments to this Agreement shall be adopted by a consensus of the Contracting Parties.

3) Acceptance of amendments shall be notified to the Depositary in writing and shall enter into force on the thirtieth day following the receipt by the Depositary of notification of the acceptance by all the Contracting Parties.

4) After the entry into force of an amendment to this Agreement any new Contracting Party to this Agreement shall become a Contracting Party to this Agreement as amended.

Article 26

Appendices and Amendments to Appendices

1) Appendices to this Agreement shall form an integral part of the Agreement.

2) Amendments to an Appendix:

   (a) Any Contracting Party may propose amendments to an Appendix at a meeting of the Contracting Parties;

   (b) Such amendments shall be adopted by a consensus of the Contracting Parties;

(c) The Depositary shall without delay communicate the amendment so adopted to all Contracting Parties.

3) The adoption and entry into force of a new Appendix to this Agreement shall be subject to the same procedure as for the adoption and entry into force of an amendment to an Appendix as provided for in paragraph (2) of this Article, Provided that the new Appendix shall not enter into force until such time as the amendment to the Agreement enters into force.

Article 27

Rules of Procedure

The Contracting Parties shall adopt rules of procedure for their meetings.

Article 28

Reports

The Contracting Parties shall transmit to the Secretariat reports on the measures adopted in implementation of this Agreement in such form and at such intervals as the meetings of Contracting Parties may determine.

Article 29

Relationships with Other Agreements

The provisions of this Agreement shall in no way affect the rights and obligations of any Contracting Party with regard to any existing treaty, convention or agreement.

Article 30

Settlement of Disputes

Any dispute between the Contracting Parties arising out of the interpretation or implementation of this Agreement shall be settled amicably by consultation or negotiation.

Article 31

Ratification

This Agreement shall be subject to ratification by the Contracting Parties. The instruments of Ratification shall be deposited with the Secretary-General of the ASEAN Secretariat, who shall assume the functions of Depositary.

Article 32

Accession

1) After the entry into force of the Agreement, any Member State may accede to this Agreement, subject to prior approval by the Contracting Parties to this Agreement.

2) Instrument of accession shall be deposited with the Depositary.

Article 33

Entry into Force

1) This Agreement shall enter into force on the thirtieth day after the deposit of the sixth Instruments of Ratification.

2) Thereafter, this Agreement shall enter into force with respect to any Contracting Party on the thirtieth day following the date of deposit of the instrument of accession by that Contracting Party.

Article 34

Responsibility of the Depositary

The Depositary shall inform the Governments which have signed this Agreement;

(a) of the deposit of instruments of ratification, acceptance or accession;

(b) of the date on which the Agreement will come into force.

Article 35

Deposit and Regulation

1) The present Agreement shall be deposited with the Depositary who shall transmit certified true copies thereof to the Governments of all Contracting Parties which have signed the present Agreement or acceded to it.

2) As soon as the present Agreement enters into force, the text shall be transmitted by the Depositary to the Secretary-General of the United Nations for registration and publication, in accordance with Article 102 of the Charter of the United Nations.

16

International Plant Protection Convention

Preamble

The Contracting Governments, recognizing the usefulness of international co-operation in controlling pests and diseases of plants and plant products and in preventing their introduction and spread across national boundaries, and desiring to ensure close co-ordination of measures directed to these ends, have agreed as follows:

Article I

Purpose and Responsibility

1. With the purpose of securing common and effective action to prevent the introduction and spread of pests and diseases of plants and plant products and to promote measures for their control, the contracting Governments undertake to adopt the legislative, technical and administrative measures specified in this Convention and in supplementary agreements pursuant to Article III.

2. Each contracting Government shall assume responsibility for the fulfilment within its territories of all requirements under this Convention.

Article II

Scope

1. For the purposes of this Convention the term "plant" shall comprise living plants and parts thereof, including seeds in so far as the supervision of their importation under Article VI of the Convention or the issue of phytosanitary certificates in respect of them under Articles IV(1), (a) (iv) and V of this Convention may be deemed necessary by contracting Governments; and the term "plant products" shall comprise unmanufactured and milled material of plant crigin, including seeds in so far as they are not included in the term "plants".

2. The provisions of this Convention may be deemed by contracting Governments to extend to storage places, containers, conveyances, packing material and accompanying media of all sorts including soil involved in the international transportation of plants and plant products

3. This Convention shall have particular reference to pests and diseases of importance to international trade.

Article III

Supplementary Agreements

1. Supplementary agreements applicable to specific regions, to specific pests and diseases, to specific plants and plant products, to specific methods of international transportation of plants and plant

products, or otherwise supplementing the provisons of this Convention, may be proposed by the Food and Agriculture Organization of the United Nations (hereinafter referred to as "FAO") on the recommendation of contracting Government or on its own initiative, to meet special problems of plant protection which need particular attention or action.

2. Any such supplementary agreements shall come into force for each contracting Government after acceptance in accordance with the provisions of the FAO Constitution and Rules of Procedure.

Article IV

National Organization for Plant Protection

1. Each contracting Government shall make provision, as soon as possible and to the best of its ability, for

   a) an official plant protection organization, with the following main functions:

      (i) the inspection of growing plants, of areas under cultivation (including fields, plantations, nurseries, gardens and greenhouses) , and of plants and plant products in storage and in transportation particularly with the object of reporting the existence, outbreak and spread of plant diseases and pests and of controlling those pests and diseases;

      (ii) the inspection of consignments of plants and plant products moving in international traffic, and, as far as practicable, the inspection of consignments of other articles or commodities moving in international traffic under conditions where they may act incidentally as carriers of pests and diseases of plants and plant products, and the inspection and supervision of storage and transportation facilities of all kinds involved in international traffic whether of plants and plant products or of other commodities, particularly with the object of preventing the dissemination across national boundaries of pests and diseases of plants and plant products;

(iii) the disinfestation or disinfection of consignments of plants and plant products moving in international traffic, and their containers, storage places, or transportation facilities of all kinds employed;

(iv) the issue of certificates relating to phytosanitary conditions and origin of consignments of plants and plant products (hereinafter referred to as "phytosanitary certificates");

b) the distribution of information within the country regarding the pests and diseases of plants and plant products and the means of their prevention and control;

c) research and investigation in the field of plant protection.

2. Each contracting Government shall submit a description of the scope of its national organization for plant protection and of changes in such organization to the Director-General of FAO, who shall circulate such information to all contracting Governments.

Article V

Phytosanitary Certificates

1. Each contracting Government shall make

arrangements for the issue of phytosanitary certificates to accord with the plant protection regulations of other contracting Governments and in conformity with the following provisions:

a) Inspection shall be carried out and certificates issued only by or under the authority of technically qualified and duly authorized officers and in such circumstances and with such knowledge and information available to those officers that the authorities of importing countries may accept such certificates with confidence as dependable documents.

b) Each certificate covering materials intended for planting or propagation shall be as worded in the Annex to this Convention and shall include such additional declarations as may be required by the importing country. The model certificate may also be used for other plants or plant products where appropriate and not inconsistent with the requirements of the importing country.

c) The certificates shall bear no alterations or erasures.

2. Each contracting Government undertakes not to require consignment of plants intended for planting or propagation imported into its territories to be accompanied by phytosanitary certificates inconsistent with the model set out in the Annex to this Convention.

Article VI

Requirements in Relation to Imports

1. With the aim of preventing the introduction of diseases and pests of plants into their territories , contracting Governments shall have full authority

to regulate the entry of plants and plant products, and to this end, may:

a) prescribe restrictions or requirements concerning the importation of plants or plant products;

b) prohibit the importation of particular plants or plant products, or of particular consignments of plants or plant products;

c) inspect or detain particular consignments of plants or plant products;

d) treat, destroy or refuse entry to particular consignments of plants or plant products, or require such consignments to be treated or destroyed.

2. In order to minimize interference with international trade, each contracting Government undertakes to carry out the provisions referred to in paragraph 1 of this Article in conformity with the following:

a) Contracting Governments shall not, under this plant protection legislation, take any of the measures specified in paragraph 1 of this Article unless such measures are made necessary by phytosanitary considerations.

b) If a contracting Government prescribes any restrictions or requirements concerning the importation of plants and plant products into its territories, it shall publish the restrictions or requirements and communicate them immediately to the plant protection services of other contracting Governments and to FAO.

c) If a contracting Government prohibits, under the provisions of its plant protection legislation, the importation of any plant or plant products, it shall publish its decision with reasons and shall immediately inform the plant protection services of other contracting Governments and FAO.

d) If a contracting Government requires consignment of particular products to be imported only through specified points of entry, such points shall be so selected as not unnecessarily to impede international commerce. The contracting Government shall publish a list of such points of entry and communicate it to the plant protection services of other contracting Governments and to FAO, such restrictions on points of entry shall not be made unless the plants or plant products concerned are required to be accompanied by phytosanitary certificates or to be submitted to inspection or treatment.

e) Any inspection by the plant protection service of a contracting Government of consignments of plants offered for importation shall take place as promptly as possible with due regard to the perishability of the plants concerned. If any consignment is found not to conform to the requirements of the plant protection legislation of the importing country, the plant protection service of the exporting country shall be informed. If the consignment is destroyed, in whole or in part, and official report shall be forwarded immediately to the plant protection service of the exporting country.

f) Contracting Governments shall make provisions which, without endangering their own plant production, will reduce to a minimum the number of cases in which a phytosanitary certificate is required on the entry of plants or plant products not intended for planting, such as cereals, fruits, vegetables and cut flowers.

g) Contracting Governments may make provision for the importation for purposes of scientific research of plants and plant products and of specimens of plant pests and disease-causing organisms under conditions affording ample precaution against the risk of spreading plant diseases and pests.

3. The measures specified in this Article shall not be applied to goods in transit throughout the territories of contracting Governments unless such measures are necessary for the protection of their own plants.

Article VII

International Co-operation

The contracting Governments shall co-operate with one another to the fullest practicable extent in achieving the aims of this Convention, in particular as follows:

a) Each contracting Government agrees to co-operate with FAO in the establishment of a world reporting service on plant diseases and pests, making full use of the facilities and services of existing organizations for this purpose, and, when this is established, to furnish to FAO periodically the following information.

(i) reports on the occurrence, outbreak and spread of economically important pests and diseases of plants and plant products which may be of immediate or potential danger;

(ii) information on means found to be

effective in controlling the pests and diseases of plants and plant products.

b) Each contracting Government shall, as far as is practicable, participate in any special campaigns for combating particular destructive pests or diseases which may seriously threaten crop production and need international action to meet the emergencies.

Article VIII

Regional Plant Protection Organization

1. The contracting Governments undertake to co-operate with one another in establishing regional plant protection organizations in appropriate areas.

2. The regional plant protection organizations shall function as the co-ordinating bodies in the areas covered and shall participate in various activities to achieve the objectives of this Convention.

Article IX

Settlement of Disputes

1. If there is any dispute regarding the interpretation or application of this Convention, or if a contracting Government considers that any action by another contracting Government is in conflict with the obligations of the latter under Articles V and VI of this Convention, especially regarding the basis of prohibiting or restricting the imports of plants or plant products coming from its territories, the Government or Governments concerned may request the Director-General of FAO to appoint a committee to consider the question in dispute.

2. The Director-General of FAO shall thereupon, after consultation with the Governments concerned, appoint a committee of experts which

shall include representatives of those Governments. This committee shall consider the question in dispute, taking into account all documents and other forms of evidence submitted by the Governments concerned. This committee shall submit a report to the Director-General of FAO who shall transmit it to the Governments concerned, and to other contracting Governments.

3. The contracting Governments agree that the recommendations of such a committee, while not binding in character, will become the basis for renewed consideration by the Governments concerned of the matter out of which the disagreement arose.

4. The Governments concerned shall share equally the expenses of the experts.

Article X

Substitution of Prior Agreements

This Convention shall terminate and replace, between contracting Governments, the International Convention respecting measures to be taken against the Phylloxera vastarix of 3 November 1881, the additional Convention signed at Berne on 15 April 1889 and the International Convention for the Protection of Plants signed at Rome on 16 April 1929.

Article XI

Territorial Application

1. Any Government may at the time of ratification or adherence or at any time thereafter communicate to the Director-General of FAO a declaration that this Convention shall extend to all or any of the territories for the international relations of which it is responsible, and this Convention shall be applicable to all territories

specified in the declaration as from the thirtieth day after the receipt of the declaration by the Director-General.

2. Any Government which has communicated to the Director-General of FAO a declaration in accordance with paragraph 1 of this Article may at any time communicate a further declaration modifying the scope of any former declaration or determinating the application of the provisions of the present Convention in respect of any territory. Such modification or termination shall take effect as from the thirtieth day after the receipt of the declaration by the Director-General.

3. The Director-General of FAO shall inform all signatory and adhering Governments of any declaration received under this Article.

Article XII

Ratification and Adherence

1. This Convention shall be open for signature by all Governments until 1 May 1952 and shall be ratified at the earliest possible date. The instruments of ratification shall be deposited with the Director-General of FAO, who shall give notice of the date of deposit to each of the signatory Governments.

2. As soon as this Convention has come into force in accordance with Article XIV, it shall be open for adherence by non-signatory Governments. Adherence shall be effected by the deposit of an instrument of adherence with the Director-General of FAO, who shall notify all signatory and adhering Governments.

Article XIII

Amendment

1. Any proposal by a contracting Government for the amendment of this Convention shall be communicated to the Director-General of FAO.

2. Any proposal amendment of this Convention received by the Director-General of FAO from a contracting Government shall be presented to a regular or special session of the Conference of FAO for approval and , if the amendment involves important technical changes or imposes additional obligations on the contracting Governments, it shall be considered by an advisory committee of specialists convened by FAO prior to the Conference.

3. Notice of any proposed amendment of this Convention shall be transmitted to the contracting Governments by the Director-General of FAO not later than the time when the agenda of the session of the Conference at which the matter is to be considered is dispatched.

4. Any such proposed amendment of this Convention shall require the approval of the Conference of FAO and shall come into force as from the thirtieth day after acceptacne by two-thirds of the contracting Governments.

5. The instruments of acceptance of amendments involving new obligations shall be deposited with the Diretor-General of FAO, who shall inform all contracting Governments of the receipt of acceptances and the entry into force of amendments.

Article XV

Denunciation

1. Any contracting Government may at any time give notice of denunciation of this Convention by notificaion addressed to the Director-General of FAO. The Director-General shall at once inform all signatory and adhering Governments.

2. Denunciation shall take effect one year form the date of receipt of the notification by the Director-General of FAO.

17

Convention on Biological Diversity

Biological diversity is being significantly reduced by certain human activities. Aware of the general lack of information and knowledge regarding biological diversity and of the urgent need to develop scientific, technical and institutional capacities to provide the basic understanding upon which to plan and implement appropriate measures.

Noting that it is vital to anticipate, prevent and attack the causes of significant reduction or loss of biological diversity at source. Noting also that whether there is a threat significant reduction or loss of biological diversity. lack of full scientific certainty should not be used as a reason for postponing measures to avoid or minimise such a threat. Noting further that the fundamental requirement for the conservation of biological diversity is the *in situ*, conservation of ecosystems and natural habitats and the maintenance and recovery of viable populations of species in their natural surroundings.

Noting further that *ex situ* measures, preferably in the country of origin, also have an important role to play. Recognising the close and traditional dependence of any indigenous and local communities embodying traditional lifestyles on biological resources, and the desirability of sharing equitably benefits arising from the use of traditional knowledge, innovations and practices relevant to the conservation of biological diversity and the sustainable use of its components. Recognising also the vital role that women play in the conservation and sustainable use of biological diversity and affirming the need for the full participation of women at all levels of policy-making and implementation for biological diversity conservation.

Stressing the importance of, and the need to promote, international, regional and global cooperation among States and intergovernmental organisations and the non governmental sector for the conservation of biological diversity and the sustainable use of its components. Acknowledging further that special provision of new and additional financial resources and appropriate access to relevant technologies can be expected to make a substantial difference in the world's ability to address the loss of biological diversity.

Acknowledging further that special provision is required to meet the needs of developing countries, including the provision of new and additional financial resources and appropriate access to relevant technologies.

Noting in this regard the special conditions of the least developed countries and small island States. Acknowledging that substantial investments are required to conserve biological diversity and that there is the expectation of a broad range of environmental, economic and social benefits from those investments. Recognising that economic and social development and poverty eradication are the first and overriding priorities of developing countries.

Aware that conservation and sustainable use of biological diversity is of critical importance for meeting the food, health and other needs of the growing world population, for which purpose access to and sharing of both genetic resources and technologies are essential. Noting that, ultimately, the conservation and sustainable use of biological diversity will strengthen friendly relations among States and contribute to peace for humankind.

Desiring to enhance and complement existing international arrangement for the conservation of biological diversity and sustainable use of its components, and Determined to conserve and sustainable use biological diversity for the benefit of present and future generations.

Have agreed as follows:

Article 1. Objectives

The objectives of this Convention, to be pursued in accordance with its relevant provisions, are the conservation of biological diversity, the sustainable use of its components and the fair and equitable sharing of the benefits arising out of the utilisation of genetic resources, including by appropriate access to genetic resources and by appropriate transfer of relevant technologies, taking into account all rights over those resources and to technologies, and by appropriate funding.

Article 2. Use of Terms

For the purposes of this Convention:

"Biological diversity" means the variability among living organisms from all sources, including, inter alia, terrestrial, marine and other aquatic ecosystems and the ecological complexes of which they are part; this includes diversity within, species, between species and of ecosystems.

"Biological resources" includes genetic resources, organisms or parts thereof, populations, or any other biotic component of ecosystems with actual or potential use or value for humanity. "Biotechnology" means any technological application that uses biological systems, living organisms, so derivatives therefore , to make or modify products or process for specific use.

"Country of origin of genetic resources" means the country which possesses those genetic resources is in situ conditions. "Country providing genetic resources" means the country supplying genetic resources collected from in situ sources, including populations of both wild and domesticated species, or taken from ex situ sources, which may or may not have originated in that country "Domesticated or cultivated species" means species in which the evolutionary process has been influenced by humans to meet their needs.

"Ecosystem" means a dynamic complex of plant, animal and microorganism communities and their nonliving environment interacting as a functional unit.

"*Ex situ*, conservation" means the conservation of components of biological diversity outside their natural habitats.

"Genetic Material" means any material of plant, animal, microbial or other origin containing functional units of heredity.

"Genetic resources" means genetic material of actual or potential value.

"Habitat" means the place or type of site where an organism or population naturally occurs.

"*In situ* conditions" means conditions where genetic resources exist within ecosystems and natural habitats and the maintenance and recovery of viable populations of species ins their natural; surrounding and, in the case, of domesticated or cultivated species, in the surrounding where they have developed their distinctive properties.

"Protected area" means a geographically defined area which is designated or regulated and managed to achieve specific conservation objectives.

"Regional economic integration organisation" means an organisation constituted by sovereign States of a given region, to which its member States have transferred competence in respect of matters governed by this Convention and which has been duly authorised, an accordance with its internal procedures, to sign, ratify, accept, approve or accede to it.

"Sustainable use" The use of components of biological diversity in a way and at a rate that does not lead to long term decline of biological diversity, thereby maintaining its potential to meet the needs and aspirations of present and future generations.

"Technology" includes biotechnology.

Article 3. Principle

States have, in accordance with the Charter of the United Nations and the principles of international law, the sovereign right to exploit their own resources pursuant to their own environmental policies, and the responsibility to ensure that activities within their jurisdiction or control for not cause damage to the environment of other States or of areas beyond the limits of national Jurisdiction.

Article 4. Jurisdictional Scope

Subject to the rights of other States, and except as otherwise expressly provided in this Convention, the provisions of this Convention apply, in relation to each Contracting Party:

a. In the case of components of biological diversity, in areas within the limits of its national jurisdiction; and

b. In the case of processes and activities, regardless of where their effects occur, carried out under its jurisdiction or control, within the area of its national jurisdiction or beyond the limits of national jurisdiction.

Article 5. Cooperation

Each Contracting Party shall, as far as possible and as appropriate, cooperate with other contracting parties, directly or, where appropriate, through competent international organisations in respect of areas beyond national jurisdiction and on other matters of mutual interest, for the conservation and sustainable use of biological diversity.

Article 6. General Measures for Conservation and Sustainable Use

Each Contracting Party shall, in accordance with its particular conditions and capabilities:

a. Develop national strategies, plans or programmes for the conservation and sustainable use of biological diversity or adapt for this purpose existing strategies, plans or programmes which shall reflect, inter alia, the measures set out in this Convention relevant to the Contracting Party concerned; and

b. Integrate as far as possible and as appropriate, the conservation and sustainable use of biological diversity into relevant sectoral or cross-sectoral plans, programmes and policies.

Article 7. Identification and Monitoring

Each Contracting Party shall, as far as possible and as appropriate, in particular for the purposes of Articles 8 to 10:

a. Identify components of biological diversity important for its conservation and sustainable use having regard to the indicative list of categories set down in Annex I:

b. Monitor, through sampling and other techniques, the components of biological diversity identified pursuant to subparagraph (a) above, paying particular attention to those requiring urgent conservation measures and those which offer the greatest potential for sustainable use;

Identify process and categories of activities which have or are likely to have significant adverse impacts on the conservation and sustainable use of biological diversity, and monitor their effects through sampling and other techniques: and Maintain and organise by any mechanism data derived from identification and monitoring activities pursuant to subparagraphs (a), (b) and (c) above.

Article 8. *In situ* Conservation

Each Contracting Party shall, as far as possible and as appropriate:

a. Establish a system of protected areas or areas where special measures need to be taken to conserve biological diversity;

b. Develop, where necessary, guidelines for the selection, establishment and management of protected areas where special measures need to be taken to conserve biological diversity;

c. Regulate or manage biological resources for the conservation of biological diversity within or outside protected areas with a view to ensuring their conservation and sustainable use;

d. Promote the protection of ecosystems, natural habitats and maintenance of viable populations of species in natural surroundings;

e. Promote environmentally sound and sustainable development in areas adjacent to protected areas with a view to furthering protection of these areas;

f. Rehabilitate and restore degraded ecosystems and promote the recovery of threatended species, inter alia, through the development and implementation of plans or other management strategies;

g. Establish or maintain means to regulate, manage or control the risks associated with the use and release of living modified organisms resulting from biotechnology which are likely to have adverse environmental impacts that could affect the conservation and sustainable use of biological diversity, taking also into account the risks to human health;

h. Prevent the introduction of, control or eradicate those alien species which threaten ecosystems, habitats or species;

i. Endeavour to provide the conditions needed or camptibility between present uses and the conservation of biological diversity and the sustainable use of its components

j. Subject to its national legislation, respect, preserve and maintain knowledge, innovations and practices of indigenous and local communities embodying traditional lifestyles relevant for the conservation and ustinable use of biological diversity and promote their wider application with the approval and involvement of the holders of such knowledge, innovations and practices and encourage the equitable sharing of the benefits arising from the utilisation of such knowledge, innovations and practices;

k. Develop or maintain necessary legislation and/or other regulatory provisions for the protection of threatened species and populations;

l. Where a significant adverse effect on biological diversity has been determined pursuant to Article 7, regulate or manage the relevant processes and categories of activities; and

m. Cooperate in providing financial and other support for in situ conservation outlined in subparagraphs (a) to (1) above, particularly to developing countries.

Article 9. Ex Situ Conservation

Each Contracting Party shall, as far as possible and as appropriate, and predominantly for the purpose of complementing in situ measures.

a. Components of biological diversity, of such components;

b. Conservation of and research on plants, animals and microorganisms;

c. Adopt measures for the recovery and rehabilitation of threatened species and for their reintroduction into appropriate conditions;

d. Regulate and manage collection of biological resources from purposes so as not to threaten ecosystem and in situ population of species except where special temporary ex situ measures are required under subparagraph (c) above; and

e. Cooperate in providing for ex situ conservation outlined in subparagraphs (a) to (d) above and in the establishment and maintenance of ex situ conservation facilities in developing countries.

Article 10: Sustainable use of components of biological diversity

Each Contracting Party shall, as far as possible and as appropriate

a. Integrate Consideration of the conservation and sustainable use of biological resources;

b. Adopt measures relating to the use of biological resources to avoid or minimise adverse impacts on biological diversity;

c. Protect and encourage customary use of biological resources to avoid or minimise adverse impacts on biological diversity;

d. Support local populations to develop and implement remedial action in degraded areas where biological diversity has been reduced: and

e. Encourage cooperation between its governmental authorities and its private sector in developing methods for sustainable use of biological resources.

Article 11. Incentive Measures

Each Contracting Party shall, as far as possible and as appropriate, adopt economically and socially sound measures that act as incentives for the conservation and sustainable use of components of biological diversity.

Article 12. Research and Training

The Contracting Parties, taking into account the special needs of developing countries, shall:

a. Establish and maintain programmes for scientific and technical education and training in measures for the identification, conservation and sustainable use of biological diversity and its components and provide support for such education and training for the specific needs of developing countries

b. Promote and encourage research which contributes to the conservation and sustainable use of biological diversity, particularly in developing countries inter alia, in accordance with decisions of the Conference of the Parties taken consequence of recommendations of the subsidiary Body on Scientific, Technical and Advice; and

c. In keeping with the provisions of Articles 10, 18 and 20, promote and cooperate in the use of scientific advances in biological diversity research in developing methods for conservation and sustainable use of biological resources.

Article 13. Public Education and Awareness

The Contracting Parties shall:

a. Promote and encourage understanding of the importance of, and the measures required for, the conservation of biological diversity, as well as its propagation through media, and the inclusion of these topics in educational programmes; and

b. Cooperate, as appropriate, with other States and international organisations in developing educational and public awareness programmes, with respect to conservation and sustainable use of biological diversity.

Article 14. Impact Assessment and Minimising Adverse Impacts

I. Each Contracting Party, as far as possible and as appropriate, shall:

a. Introduce appropriate procedures requiring environmental impact assessment of its proposed projects that are likely to have significant adverse effects on biological diversity with a view to avoiding or minimising such effects and, where appropriate, allow for public participation in such procedures.

b. Introduce appropriate arrangements to ensure that the environmental consequence, of its programmes and policies that are likely to have significant adverse impacts on biological diversity are duly taken into account;

c. Promote, on the basis of reciprocity, notification, exchange of information and consultation on activities under their jurisdiction or control which are likely significantly to affect adversely the biological diversity of other States or areas beyond the limits of national jurisdiction, by encouraging the conclusion of bilateral, regional or multilateral arrangements as appropriate;

d. In the case of imminent or grave danger or damage originating under its jurisdiction or control to biological diversity within the area under jurisdiction of other States or in areas beyond the limits of national jurisdiction of other

States or in areas beyond the limits of national jurisdiction, notify immediately the potentially affected States of such danger or damage, as well as initiate action to prevent or minimise such danger or damage:

e. Promote national arrangements for emergency responses to activities or events, where caused naturally or otherwise, which present a grave and imminent danger to biological diversity and encourage international cooperation to supplement such national efforts and, where appropriate and agreed by the States or regional economic integration organisations concerned, to establish joint contingency plans.

2. The Conference of the Parties shall examine, on the basis of studies to be carried out, the issue of liability and redress, including restoration and compensation, for damage to biological diversity, except where such liability is a purely internal matter.

Article 15. Access to Genetic Resources

1. Recognising the sovereign rights of States over their natural resources, the authority to determine access to genetic resources rests with the national governments and is subject to national legislation.

2. Each Contracting Party shall endeavour to create conditions to facilitate access to genetic resources for environmentally sound uses by other Contracting Parties and to impose restrictions that run counter to the objectives of this Convention.

3. For the purpose of this Convention, the genetic resources being provided by a Contracting Party, as referred to in this Article and Article 16 and

19. origin of such resources or by the Parties that have acquired the genetic resources in complete with this Convention.

4. Access, where granted, shall be on mutually agreed terms and subject to the provisions of the Article.

5. Access to genetic resources shall be subject to prior informed consent of the Contracting Party providing such resources, unless otherwise determined by that Party.

6. Each Contracting Party shall endeavour to develop and carry out scientific research based on genetic resources provided by other Contracting Parties with the full participation of, and where possible in, such Contracting Parties.

7. Each Contracting Party shall take legislative, administrative or policy measures, as appropriate, and in accordance with Articles 16 and 19 and where necessary through the financial mechanism established by Articles 20 ant 21, with the aim of starting in a fair and equitable way the results of research and development and the benefits arising from the commercial and other utilisation of genetic resources with the Contracting Party providing such resources. Such sharing shall be upon mutually agreed terms.

Article 16. Access to and Transfer of Technology

1. Each Contracting Party, recognising that technology includes biotechnology, and that both access to and transfer of technology among Contracting Parties are essential elements for the attainment of the objectives of this Convention, undertakes subjects to the provisions of this Article to provide and/or facilitate access for an

transfer to other Contracting Parties of technologies that are relevant to the conservation and sustainable use of biological diversity or make use of genetic resources and do not cause significant damage to the environment.

2. Access to and transfer of technology referred to in paragraph 1 above to developing countries shall be provided and/or facilitated under fair and most favourable terms, including on concessional and preferential terms where agreed, and where necessary in accordance with the financial mechanism established by Articles 20 and 21. In the case of technology to patents and other intellectual property rights, such access and transfer shall be provided on terms with recognise and are consistent with the adequate and effective protection of intellectual property rights. The application of this paragraph shall he consistent with paragraphs 3,4 and 5 below.

3. Each Contracting Party shall take legislative, administrative or policy measures, as appropriate, with the aim that Contracting Parties, in particular those that are developing countries, which provide genetic resources are provided access to and transfer of technology which makes use of those resources,, on mutually agreed terms, including technology protected by patents and other intellectual property rights, where necessary through the provisions of Articles 20 and 21 and in accordance with international law and consistent with paragraphs 4 and 5 below.

4. Each Contracting Party shall take legislative, administrative or policy measures, as appropriate, with the aim that the private sector facilitates access to, joint development and transfer of technology referred to in paragraph 1 above for the benefit of both governmental institutions and

the private sector of developing countries and in this regard shall abide by the obligations included in paragraphs 1, 2, and 3 above.

5. The Contracting Parties, recognising that patents and other intellectual property rights may have an influence on the implementation of this Convention, shall cooperate in this regard subject to national legislation and international law in order to ensure that such rights are supportive of and do not run counter to its objectives.

Article 17. Exchange of Information

1. The Contracting Parties shall facilitate the exchange of information from all publicly available source, relevant to the conservation and sustainable use of biological diversity, taking into account the special needs of developing countries.

2. Such exchange of information shall include exchange of results of technical, scientific and socioeconomic research, as well as information on training and surveying programmes, specialised knowledged, indigenous and traditional knowledge as such and in combination with the technologies referred to in Article 16, paragraph 1. It shall also, where feasible, include repatriation of information.

Article 18. Technical and Scientific Cooperation

1. The Contracting Parties shall promote international technical and scientific cooperation in the field of conservation and sustainable use of biological diversity, where necessary through the appropriate international and national institutions.

2. Each Contracting Party shall promote technical in particular developing countries, in implementing this Convention, inter alia, through the development and implementation of national policies. In promoting such cooperation, special attention should be given to the development and strengthening of national capabilities by means of human resources development and institution building.

3. The Conference of the Parties, at its first meeting, shall determine how to establish a clearinghouse mechanism to promote and facilitate technical and scientific cooperation.

4. The Contracting Parties shall, in accordance with national legislation and policies, encourage and develop methods of cooperation for the development and use of technologies, including indigenous and traditional technologies, in pursuance of the objectives of this Convention. For this purpose, the Contracting Parties shall also promote cooperation in the training of personnel and exchange of experts.

5. The Contracting Parties shall, subject to mutual agreement, promote the establishment of joint research programmes and joint ventures for the development of technologies relevant to the objectives of this Convention.

Article 19. Handling of Biotechnology and Distribution of its Benefits

1. Each Contracting Party shall take legislative, administrative or policy measures, as appropriate, to provide for the effective participation in biotechnological research activities by those Contracting Parties, especially developing

countries, which provide the genetic resources for such research, and where feasible in such Contracting Parties.

2. Each Contracting Party shall take all practicable measures to promote and advance priority access on a fair and equitable basis by Contracting Parties, especially developing countries, to the results and benefits arising from biotechnologies based upon genetic resources provided by those Contracting Parties. Such access shall be on mutually agreed terms.

3. The Parties shall consider the need for and modalities of a protocol setting out appropriate procedures, including, in particular, advance informed agreement, in the field of the safe transfer, handling and use of any living modified organism resulting from biotechnology that may have adverse effect on the conservation and sustainable use of biological diversity.

4. Each Contracting Party shall, directly or by requiring any natural or legal person under its jurisdiction providing the organisms referred to in paragraph 3 above, provide any available information about the use and safety regulations required by that Contracting Party in handling such organisms as well as any available information on the potential adverse impact of the specific organisms concerned to the Contracting Party into which those organisms are to be introduced.

Article 20. Financial Resources

1. Each Contracting Party undertakes to provide, in accordance with its capabilities, financial support and incentives in respect of those national activities which are intended to achieve the

objectives of this Convention, in accordance with its national plans, priorities and programmes.

2. The developed country Parties shall provide new and additional financial resources to enable developing country Parties to meet the agreed full incremental costs to them of implementing measures which fulfil the obligations of this Convention and to benefit from Its provisions and which costs are agreed between a developing country Party and the institutional structure, referred to in Article 21, in accordance with policy, strategy, programme priorities and eligibility criteria and an indicate list of incremental costs established by the Conference of the Parties. Other Parties, including countries undergoing the process of transition to a market economy, may voluntarily assume the obligations of the developed country Parties. For the purpose of this Article, the Conference of' the Parties shall at its first meeting establish a list of developed country Parties and other Parties which voluntarily assume the obligations of the developed country Parties.

   The Conference of the Parties shall periodically review and if necessary amend the list. Contributions from other countries and source on a voluntary basis would also be encouraged. The implementation of these commitments shall take into account the need for adequacy, predictability and timely flow of funds and the importance of burden-sharing among the contributing Parties included in the list.

3. The developed country Parties may also provide, and developing country Parties avail themselves of, financial resources related to the implementation of this Convention through bilateral, regional and other multilateral channels.

4. The extent to which developing country Parties will effectively implement their commitments under the Convention will depend on the effective implementation by developed country Parties of their commitments sunder the Convention related to financial resources and transfer of technology and will take fully into account the fact that economic and social development and eradication of poverty are the first and overriding priorities of the developing country Parties.

5. The Parties shall take full account of the specific needs and special situation of least developed countries in their actions with regard to funding and transfer of technology.

6. The Contracting Parties shall also take into consideration the special conditions resulting from the dependence on distribution and location of biological diversity within developing country Parties, in particular small island States.

7. Consideration shall also be given to the special situation of developing countries, including those that are most environmentally vulnerable, such as those with arid and semi-arid zones, coastal and mountainous areas.

Article 21. Financial Mechanism

1. There shall be a mechanism for the provision of financial resources to developing country Parties for purposes of this Convention on a grant or concessional basis the essential elements of which are described in this Article. The mechanism shall function under the authority and guidance of, and be accountable to, the Conference of the Parties for purposes of this Convention. The operations of mechanism shall

be carried out by such institutional structure as may be decided upon by the Conference of the Parties at its first meeting. For purposes of this Convention, the Conference of the Parties shall determine the policy, strategy, programme priorities and eligibility criteria relating to the access to and utilisation of such resources. The contributions shall be such as to take into account the need for predictability, adequacy and timely flow of funds referred to in Article 20 in accordance with the amount of resources needed to be decided periodically by the Conference of the Parties included in the list referred to in Article 20, paragraph 2. Voluntary contributions may also be made by the developed country parties and by other countries and sources. The mechanism shall operate within a democratic and transparent system of governance.

2. Pursuant to the objectives of this Convention, the Conference of the Parties shall at its first meeting determine the policy, strategy and programme priorities, as well as detailed criteria and guidelines for eligibility for access to and utilisation of the financial resources including monitoring and evaluation on a regular of such utilisation. The Conference of the Parties shall decide on the arrangements to give effect to paragraph 1 above after consultation with the institutional structure entrusted with the operation of the financial mechanism.

3. The Conference of the Parties shall review the effectiveness of the mechanism established under this Article, including the criteria and guidelines referred to in paragraph 2 above, not less than two years after the entry into force of this Convention and thereafter on a regular basis. Based on such review, it shall take appropriate

action to improve the effectiveness of the mechanism if necessary.

4. The Contracting Parties shall consider strengthening existing financial institutions to provide financial resources for the conservation and sustainable use of biological diversity.

Article 22. Relationship with other International Conventions

1. The provisions of the present Convention shall not affect the rights and obligations of any Contracting Party deriving from any existing international agreement, except where the exercise of those rights and obligations would cause serious damage or a threat to biological diversity.

2. Contracting Parties shall implement this Convention with respect to the marine environment consistently with the rights and obligations of States under the law of the sea.

Article 23. Conference of the Parties

1. A Conference of the Parties is hereby established. The first meeting of the Conference of the Parties shall be convened by the Executive Director of the United Nations Environment Programme not later than one year after the entry into force of this Convention. Thereafter, ordinary meetings of the Conference of the Parties shall be held at regular intervals to be determined by the Conference at its first meeting.

2. Extraordinary meetings of the Conference of the Parties shall be held at such other times as may be deemed necessary by the Conference or at the written request of any Party, provided that within

six months of the request being communicated to them by the Secretariat, it is supported by at least one-third of the Parties.

3. The Conference of the Parties shall by consensus agree upon and adopt rules of procedure for itself and for any subsidiary body it may establish, as well as financial rules governing the funding of the Secretariat. At each ordinary meeting, it shall adopt a budget for the financial period until the next ordinary meeting.

4. The Conference of the Parties shall keep under review the implementation of the present Convention, and, for this purpose, shall:

   a. Establish the form and the intervals for transmitting the information to be submitted in accordance with Article 26 and consider such information as well as reports submitted by any subsidiary body;

   b. Review scientific, technical and technological advice on biological diversity provided in accordance with Article 25;

   c. Consider and adopt, as required, protocols in accordance with Article 28;

   d. Consider and adopt, as required, in accordance with Articles 29 and 30, amendments to this Convention and its annexes;

   e. Consider amendments to any protocol, as well as to any annexes thereto, and, if so decided, recommend their adoption to the parties to the protocol concerned;

   f. Consider and adopt, as required, in accordance with Article 30, additional annexes to this Convention;

   g. Establish such subsidiary bodies,

particularly to provide scientific and technical advice, as are deemed necessary for the implementation of this Convention;

h. Contact, through the Secretariat, the executive bodies of conventions dealing with matters covered by this Convention with a view to establishing appropriate forms of cooperation with them; and

i. Consider and undertake any additional action that may be required for the achievement of the purposes of this Convention in the light of experience gained in its operation.

5. The United Nations, its specialised agencies and the International Atomic Energy Agency, as well as any State not party to this Convention, may be represented as observers at meetings of the Conference of the Parties. Any other body or agency, whether governmental or non-governmental, qualified in fields relating to conservation and sustainable use of biological diversity which has informed the Secretariat of its wish to be represented as an observer at a meeting of the Conference of the Parties present object.

Article 24. Secretariat

1. A secretariat is hereby established. Its functions shall be:

   a. To arrange for and service meetings of the Conference of the Parties provided for in Article 23;

   b. To perform the functions assigned to it by any protocol;

   c. To prepare report on the execution of its

function under this Convention and present them to the Conference of the Parties;

d. To coordinate with other relevant international bodies, and in particular to enter into such administrative and contractual arrangements as may be required for the effective discharge of its functions; and

e. To perform such other functions as may be determined by the Conference of the Parties.

2. At its first ordinary meeting, the Conference of the Parties shall designate the secretariat from amongst those existing competent international organisations which have signified their willingness to carry out the secretariat functions under this Convention.

Article 25. Subsidiary body on scientific, technical and technological advice

1. A subsidiary body for the provision of scientific, technical and technological advice is hereby established to provide the Conference of the Parties and, as appropriate, its other subsidiary bodies with timely advice relating to the implementation of this Convention. This body shall be open to participation by all Parties and shall be multi disciplinary. It shall comprise government representatives competent in the relevant field of expertise. It shall report regularly to the Conference of the Parties on all aspects of its work.

2. Under the authority of and in accordance with guidelines laid down by the Conference of the parties, and upon its request, this body shall;

   a. Provide scientific and technical assessments of the status of biological diversity:

b. Prepare scientific and technical assessments of the effects of types of measures taken in accordance with the provisions of this Convention

c. Identity innovative, efficient and state of art technologies and know-how relating to the conservation and sustainable use of biological diversity and advise on the ways and means of promoting development and/or transfering such technologies;

d. Provide advice on scientific programmes and international cooperation in research and development related to conservation and sustainable use of biological diversity; and

e. Respond to scientific, technical, technological and methodological questions that the Conference of the Parties and its subsidiary bodies may put to the body.

The functions terms of reference, organisation and operation of this body may be further elaborated by the conference of the parties.

Article 26. Reports

Each Contracting Party shall, at intervals to be determined by the Conference of the Parties, present to the Conference of the Parties reports on measures which it has taken for the implementation of the provisions of this Convention and their effectiveness in meeting the objectives of this Convention

Article 27. Settlement of Disputes

1. In the event of a dispute between Contracting Parties concerning the interpretation or application of this Convention, the Parties concerned shall seek solution by negotiation.

2. If the parties concerned cannot reach agreement by negotiation, they may jointly seek the good offices of, or request mediation by, a third party.

3. When ratifying, accepting, approving or acceding to this Convention, or at any time thereafter, a State or regional economic integration organisation may declare in writing to the Depositary that for a dispute not resolved in accordance with paragraph 1 or paragraph 2 above, it accepts one or both of the following means of dispute, settlement as compulsory:

4. If the parties to the dispute have not in accordance with paragraph 3 above, accepted the same or any procedure, the dispute shall be submitted to conciliation in accordance with Part 2 of Annex 11 unless the parties otherwise agree.

5. The provisions of this Article shall apply with respect to any protocol except as otherwise provided in the protocol concerned.

Article 28. Adoption of Protocols

1. The Contracting Parties shall cooperate in the formulation and adoption of protocols to this Convention.

2. Protocols shall be adopted at a meeting of the Conference of the Parties.

3. The text of any proposed protocol shall be communicated to the Contracting Parties by the Secretariat at least six months before such a meeting.

Article 29. Amendment of the Convention or protocols

1. Amendments to this Convention may be proposed

by any Contracting Party. Amendments to any protocol may be proposed by any Party to that protocol.

2. Amendments to this Convention shall be adopted at a meeting of the Conference of the Parties. Amendments to any protocol shall be adopted at meeting of the Parties to the Protocol in question. The text of any proposed amendment to this Convention or to any protocol, except as may otherwise be provided in such protocol, shall be communicated to the parties to the instrument in question by the secretariat at least six months before the meeting at which it is proposed for adoption. The secretariat shall also communicate proposed amendments to the signatories to this Convention for information.

3 The Parties shall make every effort to reach agreement on any proposed amendment to this Convention or to any protocol by consensus. If all efforts at consensus have been exhausted, and no agreement reached, the amendment shall as a last resort be adopted by a two-thirds majority vote of the Parties to the instrument in question present and voting at the meeting, and shall be submitted by the Depositary to all Parties for ratification, acceptance or approval.

4. Ratification, acceptance or approval of amendments shall be notified to the Depositary in writing. Amendments adopted in accordance with paragraph 3 above shall enter into force among Parties having accepted them on the ninetieth day after the deposit of instruments of ratification, acceptance or approval by at least two-thirds of the Contracting Parties to this Convention or of the Parties to the protocol concerned, except as may otherwise be provided in such protocol. Thereafter the amendments

shall enter into force for any other Party on the ninetieth day after that Party deposits its instruments of ratification, acceptance or approval of the amendments.

5. For the purposes of this Article, "Parties present and voting" means Parties present and casting an affirmative or negative vote.

Article 30. Adoption and Amendment of Annexes

1. The annexes to this Convention or to any protocol shall form an integral part of this Convention or of such protocol, as the case may be, and, unless expressly provided otherwise, a reference to this Convention or its protocols constitutes at the same time a reference to any annexes thereto. Such annexes shall be restricted to procedural, scientific, technical and administrative matters.

2. Except as may be otherwise provided in any protocol with respect to its annexes, the following procedure shall apply to the proposal, adoption and entry into force of additional annexes to this Convention or of annexes to any protocol:

   a. Annexes to this Convention or to any protocol shall be proposed and adopted according to the procedure laid down in Article 29

   b. Any Party that is unable to approve an additional annex to this Convention or an annex to any protocol to which it is party shall so notify the Depositary, in writing, within one year from the date of the communication of the adoption by the Depositary. The Depositary shall without delay notify all Parties of any such notification received. A Party may at any time withdraw a previous declaration of objection and the annexes shall thereupon enter into force for that Party subject to subparagraph (c) below;

On the expiry of one year from the date of the communication of the adoption by the Depositary, the annex shall enter into force for all parties to this Convention or to any protocol concerned which have not submitted a notification in accordance with the provisions of subparagraph (b) above.

3. The proposal, adoption and entry into force of amendments to annexes to this Convention or to any protocol shall be subject to the same procedure as for the proposal, adoption and entry into force annexes to the Convention or annexes to any protocol.

4. If an additional annex or an amendment to an annex is related an amendment to this Convention or to any protocol, the additional annex or amendment shall not enter into force until such time as the amendment to this Convention or to the protocol concerned enters into force.

Article 31. Right to Vote

1. Except as provided for in paragraph 2 below, each Contracting Party to this Convention or to any protocol shall have one vote.

2. Regional economic integration organisations, in matters within them competence, shall exercise their right to vote if their member States which are Contracting Parties to the Convention of the relevant protocol. Such organisations shall not exercise their right to vote in their member States exercise theirs, and vice versa.

Article 32. Relationship between this Convention and its Protocols

1. A State or a regional economic integration organisation may not become a party to a protocol unless it is, or becomes at the same time a Contracting Party to this Convention.

2. Decisions under any protocol shall be taken only by the Parties to the protocol concerned. Any Contracting Party that has not ratified accepted or approved a protocol may participate as an observer in any meeting of the parties to that protocol.

Article 33. Signature

This Convention shall be open for signature at Rio de Janeiro by all States and any regional economic integration organisation from the fifth day of June 1992 until.

Article 34. Ratification, Acceptance or Approval

1. This Convention and any protocol shall be subject to ratification acceptance or approval by States and by regional economic integration organisations. Instruments of ratification, acceptance or approval shall be deposited with the Depository.

2. Any organisation referred to in paragraph 1 above which becomes a Contracting Party to this Convention or any protocol without any or its member States being a Contracting Party shall be bound by all the obligations under the Convention or the protocol, as the case may be. In the case of such organisations, one or more of whose member States is a Contracting Party to

the Convention or relevant protocol the organisation and its member States shall decide on their respective responsibilities for the performance of their obligations under the Convention or protocol, as the case may be in such cases, the organisation and the member States shall not be entitled to exercise rights under the Convention or relevant protocol concurrently.

3. In their instruments of ratification, acceptance or approval, the organisations referred to in paragraph 1 above shall declare the extent of their competence with respect to the matters governed by the Convention or the relevant protocol. These organisations shall also inform the Depositary of any relevant modification in the extent of their competence.

Article 35. Accession

1. This Convention and any protocol shall be open for accession by States and by regional economic integration organisations from the date on which the Convention or the protocol concerned is closed for signature. The instruments of accessions shall be deposited with the Depositary.

2. In their instruments of accession, the organisations referred to in paragraph 1 shall declare the extent of their competence with respect to the matters governed by the Convention or the relevant protocol. These organisations shall also inform the Depositary of any relevant modification in the extent of their competence.

3. The provisions of Article 34, paragraph 2, shall apply to regional economic integration

organisations which accede to this Convention or any protocol.

Article 36. Entry into Force

1. This Convention shall enter into force on the ninetieth day after the date of deposit of the thirtieth instrument of ratification, acceptance approval or accession.

2. Any protocol shall enter into force on the ninetieth day after the date of deposit of the number of instruments of ratification, acceptance, approval or accession, specified in that protocol, has been deposited.

3. For each Contracting Party which ratifies, accepts or approves this Convention or accedes thereto after the deposit of the thirtieth instrument of ratification, acceptance, approval or accession, it shall enter into force on the ninetieth day after the date of deposit by such Contracting Party of its instrument of ratification, acceptance, approval or accession.

4. Any protocol, except as otherwise provided in such protocol, shall enter into force for a Contracting Party that ratifies, accepts or approves that protocol or accedes thereto after its entry into force pursuant to paragraph 2 above, on the ninetieth day after the date on which that Contracting Party deposits its instrument of ratification, acceptance, approval or accession, or on the date on which the Convention enters into force for that Contracting Party, whichever shall be the later.

5. For the purposes of paragraphs 1 and 2 above, any instrument deposited by a regional economic integration organisation shall not be counted as

additional to those deposited by member States of such organisation.

Article 37. Reservations

No reservations may be made to this Convention.

Article 38. Withdrawals

1. At any time after two years from the date on which the present Convention has entered into force for a Contracting Party, that Contracting Party may withdraw from the Convention by giving written notification to the Depositary.
2. Any such withdrawal shall take place upon expiry of one Year after the date of its receipt by the Depositary, or on such later date as may be specified in the notification of the withdrawal.
3. Any Contracting Party which withdraws from this Convention shall be considered as also having withdraw from any protocol to which it is party.

Article 39. Financial Interim Arrangement

Provided that it has been fully restructured in accordance with the requirements of Article 21, the Global Environment Facility of the United Nations Development Programme, the United Nations Environment Programme and the International Bank for Reconstruction and Development shall be the institutional structure referred to in Article 21 on an interim basis, for the period between the entry into force of this Convention and the first meeting of the Conference of the Parties or until the Conference of the Parties decides which institutional structure will be designated in accordance with Article 21.

Article 40. Secretariat interim arrangements

The secretariat to be provided by the Executive Director of the United Nations Environment Programme shall be the secretariat referred to in Article 24, paragraph 2, on an interim basis for the period between the entry into force of this Convention and the first meeting of the Conference of the Parties.

Article 41. Depositary

The Secretary-General of the United Nations shall assume the functions of Depositary of this Convention and any protocols.

Article 42. Authentic Texts

1. The original of this Convention, of which the Arabic, Chinese, English, French, Russian and Spanish texts are equally authentic, shall be deposited with the Secretary-General of the United Nations.
2. In witness whereof the undersigned, being duly authorised to that effect, have signed this Convention.
3. Done at Rio de Janeiro on this fifth day of June, one thousand nine hundred and ninety-two.

Index